Environmental Assessment of Renewable Energy Conversion Technologies

Environmental Assessment of Renewable Energy Conversion Technologies

Edited by

PARIS A. FOKAIDES
School of Engineering, Frederick University, Nicosia, Cyprus

ANGELIKI KYLILI
Department of Environment, Ministry of Agriculture, Rural Development and Environment, Cyprus

PHOEBE-ZOE GEORGALI
School of Engineering, Frederick University, Nicosia, Cyprus

ELSEVIER

Elsevier
Radarweg 29, PO Box 211, 1000 AE Amsterdam, Netherlands
The Boulevard, Langford Lane, Kidlington, Oxford OX5 1GB, United Kingdom
50 Hampshire Street, 5th Floor, Cambridge, MA 02139, United States

ISBN: 978-0-12-817111-0

For Information on all Elsevier publications
visit our website at https://www.elsevier.com/books-and-journals

Publisher: Candice Janco
Acquisitions Editor: Jessica Mack
Editorial Project Manager: Alice Grant
Production Project Manager: Sruthi Satheesh
Cover Designer: Miles Hitchen

Typeset by MPS Limited, Chennai, India

Contents

Section D

List of contributors

Michael Bampaou
Chemical Process and Energy Resources Institute (CPERI), Centre for Research and Technology Hellas (CERTH), Thessaloniki, Greece

Elias Christoforou
School of Engineering and Applied Sciences, Frederick University, Nicosia, Cyprus

Paris A. Fokaides
School of Engineering, Frederick University, Nicosia, Cyprus

Guillermo Garcia-Garcia
Department of Chemical and Biological Engineering, The University of Sheffield, Sheffield, United Kingdom; Department of Agrifood System Economics, Centre 'Camino de Purchil', Institute of Agricultural and Fisheries Research and Training (IFAPA), Granada, Spain

Eskinder Gemechu
Faculty of Engineering, Department of Mechanical Engineering, University of Alberta, Edmonton, AB, Canada

Phoebe-Zoe Georgali
School of Engineering, Frederick University, Nicosia, Cyprus

Effrosyni Giama
Department of Mechanical Engineering, Aristotle University of Thessaloniki, Thessaloniki, Greece

Giorgos Kardaras
Chemical Process and Energy Resources Institute (CPERI), Centre for Research and Technology Hellas (CERTH), Thessaloniki, Greece; Department of Mechanical Engineering, University of Western Macedonia, Greece

Panagiota Konatzii
School of Engineering, Frederick University, Nicosia, Cyprus

Tzouliana Kraia
Chemical Process and Energy Resources Institute (CPERI), Centre for Research and Technology Hellas (CERTH), Thessaloniki, Greece

Amit Kumar
Faculty of Engineering, Department of Mechanical Engineering, University of Alberta, Edmonton, AB, Canada

Angeliki Kylili
Department of Environment, Ministry of Agriculture, Rural Development and Environment, Cyprus

Paola Lettieri
Department of Chemical Engineering, University College London, London, United Kingdom

M. A. Parvez Mahmud
School of Engineering, Deakin University, Geelong, VIC, Australia

Stephen McCord
Department of Chemical and Biological Engineering, The University of Sheffield, Sheffield, United Kingdom

Maria Milousi
Department of Chemical Engineering, University of Western Macedonia, Koila, Greece

Abayomi Olufemi Oni
Faculty of Engineering, Department of Mechanical Engineering, University of Alberta, Edmonton, AB, Canada

Kyriakos Panopoulos
Chemical Process and Energy Resources Institute (CPERI), Centre for Research and Technology Hellas (CERTH), Thessaloniki, Greece

Andrea Paulillo
Department of Chemical Engineering, University College London, London, United Kingdom

Md Mustafizur Rahman
Faculty of Engineering, Department of Mechanical Engineering, University of Alberta, Edmonton, AB, Canada

Holger Schlör
Forschungszentrum Jülich, Jülich, Germany

Manolis Souliotis
Department of Chemical Engineering, University of Western Macedonia, Koila, Greece

Alberto Striolo
School of Chemical, Biological and Materials Engineering, University of Oklahoma, Norman, OK, United States

Peter Styring
Department of Chemical and Biological Engineering, The University of Sheffield, Sheffield, United Kingdom

Nahin Tasmin
Department of Mechanical Engineering, Rajshahi University of Engineering & Technology, Kazla, Rajshahi, Bangladesh

Sandra Venghaus
Forschungszentrum Jülich, Jülich, Germany; School of Business and Economics, RWTH Aachen University, Aachen, Germany

About the editors

Dr.-Ing. Paris A. Fokaides is an Associate Professor at the School of Engineering of Frederick University, Cyprus, and a research mentor at Kaunas University of Technology, Lithuania. In Frederick University, Paris is lecturing the courses of Fluid Mechanics and Heat Transfer at the Department of Mechanical Engineering, as well as the courses of Sustainable Energy Resources, and Energy Design of Buildings in the Masters Programme of Energy Engineering, which he also coordinates. Paris holds a PhD from the University of Karlsruhe, in Germany in the field of Process Engineering and a Diploma in Mechanical Engineering of Aristotle University in Thessaloniki, Greece. Paris research is related to the promotion of Industry 4.0 practices for the assessment of the energy and sustainability performance of energy technologies and smart buildings, as well as the field of digitization and analysis of energy related processes. Paris leads the Sustainable Energy Research Group at Frederick University, an ISO 9001 certified self-funded research team consisting of 10 FTE researchers, involved in numerous European and national funded R&I activities. Paris is also Editor in Chief of the International Journal of Sustainable Energy, and member in numerous editorial boards of scientific journals. As of mid-22, Paris has authored and co-authored over 125 Scopus indexed studies, and has an h-index of 30.

Dr. Angeliki Kylili is an Environment Officer at the Department of Environment of the Ministry of Agriculture, Rural Development and Environment of the Republic of Cyprus. She has studied BSc Environmental Science and MSc Energy and Environment at the University of Leeds, United Kingdom, and has obtained her PhD in Civil Engineering with the Sustainable Energy Research Group (SERG) at Frederick University, Cyprus. Her research is primarily concerned with Life Cycle

Assessment and the exploitation of renewable energy sources. She is the author and co-author of 37 publications in international peer-reviewed journals and 4 book chapters, with an h-index of 20. Her current duties as an Environment Officer concern the development and effective implementation of the national and European policy framework for the protection of the environment. Angeliki is a national focal point for the Transport, Health and Environment Pan-European Programme (THE PEP) of the United Nations Economic Commission for Europe (UNECE), and she is also responsible for following through and providing relevant national contributions to the work of the United Nations Environment Programme (UNEP) and the Committee on Environmental Policy of UNECE.

Ms. Phoebe-Zoe Georgali is a Mechanical Engineer (BSc Mechanical Engineering) graduate from the Technological Education Institute of Chalkida, Greece, 2012 and Energy Engineer postgraduate (MSc Sustainable Energy Systems) at Frederick University, Cyprus, 2017. Since September 2020, Ms. Georgali is a PhD Candidate at Frederick University as a member of the Sustainable Energy Research Group (SERG), engaging with state-of-the-art research regarding sustainable and waste energy technologies, as well as LCA of products and services.

SECTION A

CHAPTER 1

Introduction: environmental assessment of renewable energy and storage technologies: current status

Panagiota Konatzii and Paris A. Fokaides
School of Engineering, Frederick University, Nicosia, Cyprus

Content

We live in an era where the term renewable energy has been linked to environmental-friendly and sustainable practices for converting natural resources to end energy. Countries and organizations around the world, one after another, set quantitative targets for promoting energy production with the use of renewable energy sources. The European Union (EU) is a pioneer in this field, with ambitious goals dating back to the early 2000s, which are currently being remarkably achieved. The infamous EU target of the triple 20 for 2020 with the reference year of 2005, that is, 20% energy savings, 20% promotion of the use of renewable energy sources, and 20% reduction of greenhouse gases (GHG), was not only achieved but gave way to a more ambitious goal for 2030 and 2050, resulting from the Green Deal (European Environmental Agency, 2021). The member states of the EU are moving fast towards achieving the ambitious goal of 55% energy savings by 2030, in accordance with the Fitfor55 policy framework (European Parliament, 2021). In addition to the ambitious European program, the United Nations is moving fast with the Sustainable Development Goals program, a scheme within which optimistic sustainability goals should be achieved in 17 areas, including green and sustainable energy, as well as sustainable cities and societies (United Nations, 2021).

Environmental Assessment of Renewable Energy Conversion Technologies
DOI: https://doi.org/10.1016/B978-0-12-817111-0.00013-9

Under these conditions, the promotion of renewable energy sources and related technologies constitutes the mainstream in the energy production field. The continuous development that prevails in the design and implementation of new renewable energy projects worldwide is accompanied by both research activities to develop more energy-efficient applications, but also environmentally smarter solutions (Christoforou and Fokaides, 2016). Inevitably, the point has been reached where the term renewable energy, in itself, is not a panacea, the answer to every solution, but should be evaluated and judged, with objective criteria (Kylili et al., 2016). A technological application, for example, for the conversion of solar energy into electricity, which requires large volumes of raw material, is not environmentally preferable, compared to another solution, which with the same degree of efficiency but with much smaller quantities of raw material, can convert the same amount of solar energy into another useful form (Souliotis et al., 2018). Therefore, the question of quantifying the environmental impact of the use of renewable energy sources reaches a point where it can no longer be answered qualitatively but needs to be substantiated, quantitative answers.

The answer to the question of how we can quantify the environmental impact of renewable energy sources is found in life cycle analysis. Life cycle analysis is a well-tested, well-established methodology that can quantify the environmental impact of any product or service throughout its life cycle. From the beginning of the 1990s, when this method appeared, until its first standardization in 1996, today, worldwide, it is considered the most comprehensive methodology for quantifying the environmental impact (Arnaoutakis et al., 2019). Since 1996 and its standardization through the ISO 14040 series standards, life cycle analysis has been the most widely used method of determining environmental impact (Christoforou et al., 2016). ISO 14040:2006 describes the principles and framework for life cycle assessment (LCA), including the definition of the goal and scope of the LCA, the life cycle inventory analysis (LCI) phase, the life cycle impact assessment (LCIA) phase, the life cycle interpretation phase, reporting and critical review of the LCA, limitations of the LCA, the relationship between the LCA phases, and conditions for the use of value choices and optional elements (EN ISO 14040, 2006). The environmental analysis of renewable energy sources is no exception in relation to the environmental burden determination practices that can be followed.

Decision-making on new installations in the field of energy production and storage using sustainable energy resources should be justified on specific quantitative parameters. Given the growing rate of installation of renewable energy and storage applications, the integral sustainability aspect of the environmental assessment should also be quantified in a similar manner to the technical and financial parameters (Fokaides and Christoforou, 2016). The recent development of comprehensive environmental assessment tools such as the life cycle assessment (LCA) and the product environmental footprint (PEF), as well as the scientific work conducted in these fields, allows for the development of a joint framework to evaluate different technologies on a common basis concerning their environmental perspectives (Pommeret et al., 2017). Despite the numerous scientific publications in this research field, a compilation of the justified knowledge in this topic is still not available for the scientific and engineering community (Christoforou and Fokaides, 2018).

Efforts to globalize the environmental assessment of services and products with the use of LCA date back to 2013. Particularly, in order to promote and establish LCA as the most common approach for the environmental assessment of services and products, the United Nations initiated in 2013 the Global Guidance on Environmental Life Cycle Impact Assessment Indicators (GLAM) initiative (United Nations Environment Program (UNEP), 2021). The aim of UNEP GLAM, under the United Nations Environmental Programme umbrella, is to improve worldwide agreement on environmental LCIA indicators, delivering tangible and specific recommendations for diverse environmental indicators and classification factors used (LCIA). The UNEP GLAM project is implemented by an international expert task force, which drafts and announces recommendations for different topic areas. Advancements are overviewed on a regular basis by expert consultation workshops and roundtable discussions organized among experts and stakeholders of the field. The UNEP GLAM experts are chosen from five different pools, which cover all interested parties in the field of LCA, including users of life cycle information, such as governmental and intergovernmental organizations, industries, NGOs, and members of the academia, life cycle thinking studies consultants, and LCIA methods and tools developers. The initiative was organized in three phases:

- In the first phase, which lasted from 2013 until 2016, specific impact categories were discussed and quantified, including GHG emissions and impacts of climate change, health impacts of fine particulate matter, human health impacts, land use related impacts on biodiversity, water use related impacts—water scarcity as well as cross-cutting issues.

- The second phase, which was implemented from 2017 until 2019, analyzed specific impact indicators, including acidification and eutrophication, land use impacts on soil quality, ecotoxicity natural resources and mineral primary resources, human toxicity as well as cross-cutting issues.
- The last phase, which started in 2019 and is still ongoing, aimed to establish a comprehensive, consistent and global environmental Life Cycle Impact Assessment Method (LCIA), building on the recommendations for nine impact categories from the first two phases.

The UNEP GLAM initiative is also supported by the Joint Research Centre (JRC) of the European Commission, at different levels, participating in meetings and providing scientific inputs, documentation, and technical support, in order to follow possible alignment with different methods' development (Joint Research Center, European Commission, 2021).

In this context, this book attempts to present the state-of-the-art in the field of environmental valuation of renewable energy sources. By gathering the opinion of the selected academics in the field of environmental valuation of renewable energy sources, this volume wishes to present the latest developments in the field. Specifically, this book hosts eleven (11) chapters, which deal with the following areas:

- Photovoltaic systems.
- Solar thermal systems for heat production.
- Wind generators.
- Thermochemical conversion of biomass into biofuels.
- Biochemical conversion of biomass into biofuels.
- Mechanical biomass processing.
- Geothermal systems.
- Hydroelectric systems.
- Hydrogen systems.
- Storage systems using batteries.

The purpose of this volume is to present a comprehensive overview of the environmental assessment of renewable energy conversion and storage technologies. This book aspires to compile the state-of-the-art in the field of the environmental assessment of renewable energy conversion and storage technologies and to deliver a common ground based on the key performance indicators for the comparative environmental evaluation of nonfossil energy sources applications. The readership of this book will have access to justified figures, approaches, and techniques for the comprehensive environmental assessment for a significant range of applications of individual sustainable energy conversion and storage technologies.

The authors of the volume mostly tried to maintain a common structure for all the chapters. Specifically, all the chapters present the theoretical background of the technology, which is examined, as well as the developments in the field. The main findings from life cycle inventories and life cycle impact assessments are then summarized, and the chapters are concluded with the main findings and future trends in the field. The volume is enriched with several diagrams, which aim at a better understanding of both the physical findings and the trends of the sector, as well as with tables that summarize the main findings of different studies in the sector.

This volume provides the state-of-the-art in both nonfossil energy conversion and storage techniques as well as in their environmental assessment. The readership will be informed about the goal and scope, the analysis boundaries, the inventory, and the impact assessment employed for the evaluation of these applications. Also, the readership will have an overview of the environmental footprint of the said technologies. This volume assembles and compiles information currently available in different sources concerning the environmental assessment of sustainable energy technologies. This feature is of important significance, as it will allow for the comparative assessments of different technologies, given specific boundary conditions such as the renewable potential and other specific features of the discussed technologies. The chapter of this volume also provides to the readership a more comprehensive overview of the entire energy supply chain, namely from production to storage, by allowing the consideration of different production and storage combinations based on their environmental assessment. The book aims to expand the boundaries of the environmental analysis of energy technologies.

This volume is intended for not only both researchers in the field of environmental assessment of renewable energy sources and for engineers in the field but also for students in the fields of environmental engineering and other relevant fields of engineering. Specifically, a nonexhaustive list of the audience to which this volume is addressed includes environmental scientists, environmental engineers, energy engineers, mechanical engineers, electrical engineers, chemical engineers, architects, and urban planners The aim of the authors of the volume was to give the general picture of the field, but also to give the impetus for new works as well as further research development.

References

Arnaoutakis, N., Milousi, M., Papaefthimiou, S., Fokaides, P.A., Caouris, Y.G., Souliotis, M., 2019. Life cycle assessment as a methodological tool for the optimum design of integrated collector storage solar water heaters. Energy 182, 1084−1099.

Christoforou, E.A., Fokaides, P.A., 2016. Life cycle assessment (LCA) of olive husk torrefaction. Renewable Energy 90, 257−266.

Christoforou, E., Fokaides, P.A., 2018. Advances in Solid Biofuels. Springer.

Christoforou, E., Fokaides, P.A., Koroneos, C.J., Recchia, L., 2016. Life Cycle Assessment of first generation energy crops in arid isolated island states: the case of Cyprus. Sustainable Energy Technologies and Assessments 14, 1−8.

EN ISO 14040 (2006). Environmental management. Life cycle assessment. Principles and framework.

European Environmental Agency (2021). Trends and Projections in Europe 2021. EEA Report No. 13/2021. Copenhagen: European Environment Agency.

European Parliament (2021). Legislative train schedule. Fit for 55 packages under the European Green Deal. <https://www.europarl.europa.eu/legislative-train/theme-a-european-green-deal/package-fit-for-55> (accessed 01.11.21.).

Fokaides, P.A., Christoforou, E., 2016. Life cycle sustainability assessment of biofuels. Handbook of Biofuels Production. Woodhead Publishing, pp. 41−60.

Joint Research Center, European Commission (2021). European platform on life cycle assessment. <https://eplca.jrc.ec.europa.eu/glam.html> (accessed 01.11.21.).

Kylili, A., Christoforou, E., Fokaides, P.A., 2016. Environmental evaluation of biomass pelleting using life cycle assessment. Biomass and Bioenergy 84, 107−117.

Pommeret, A., Yang, X., Kwan, T.H., Christoforou, E.A., Fokaides, P.A., Lin, C.S.K., 2017. Techno-economic study and environmental assessment of food waste-based biorefinery. Food Waste Reduction and Valorisation. Springer, Cham, pp. 121−146.

Souliotis, M., Panaras, G., Fokaides, P.A., Papaefthimiou, S., Kalogirou, S.A., 2018. Solar water heating for social housing: energy analysis and Life Cycle Assessment. Energy and Buildings 169, 157−171.

United Nations (2021). Sustainable development. Department of economic and social affairs. https://sdgs.un.org/goals (accessed 01.11.21.).

United Nations Environment Program (UNEP) (2021). Life cycle initiative. https://www.lifecycleinitiative.org/ (accessed 01.11.21.).

SECTION B

CHAPTER 2

Life cycle analysis of photovoltaic systems: a review

Effrosyni Giama[1] and Phoebe-Zoe Georgali[2]
[1]Department of Mechanical Engineering, Aristotle University of Thessaloniki, Thessaloniki, Greece
[2]School of Engineering, Frederick University, Nicosia, Cyprus

Contents

2.1 Introduction: European Union roadmap for energy and carbon emissions

One of the major developments of the last decade is the existence of a quite explicit regulatory framework, setting specific goals and providing supportive laws, directives, standards, methodologies, focusing on clean energy, minimizing energy consumption, and reducing CO_2 emissions. The main targets set by the European Commission in chronological order are (Giama et al., 2020):

- Target of 20–20–20 (20% improvement in energy efficiency, 20% reduction of greenhouse gas (GHG) emissions compared with the 1990s levels, and 20% increase in the share of renewable energy to at least 20% of the consumption)
- Revised target for 2030 (at least 40% reduction in GHG gas emissions compared to 1990s levels, at least 32% share for renewable energy, at least 32.5% improvement in energy efficiency)
- Next target for 2050 (85%–90% reduction of GHG gas emissions compared to 1990s levels)

Environmental Assessment of Renewable Energy Conversion Technologies
DOI: https://doi.org/10.1016/B978-0-12-817111-0.00008-5

Focusing on renewable energy sources (RES), the European Union (EU) aims to achieve a 20% share (of its final energy consumption) from RES by 2020 and at least a 32% share (not broken down into nationally binding targets) by 2030. Key instruments at the EU level to promote RES include directives, such as the 2009 Renewable Energy Directive. EU supports the legislative framework with schemes and financial programs; for instance, the EU Emission Trading Scheme is one of the EU's efforts to support RES implementation. (EurObserv, 2019). At a national level, EU states guidelines and funding programs focusing on research, development, and innovation on energy and environmental issues, such as Horizon2020. RESs are also supported through regional development funds as well as through grants and loans for RES projects and related infrastructure from the European Investment Bank and the European Fund for Strategic Investments (EurObserv, 2019) (Fig. 2.1).

It is important to discuss the share of energy related to RES. First of all, and based on the European Environment Agency statistics, RES cover solar thermal and photovoltaic (PV) energy, hydro (including tide, wave, and ocean energy), wind, geothermal energy, and all forms of biomass (including biological waste and liquid biofuels). The contribution of renewable energy from heat pumps is also covered for the member states for which this information was reported (Zampori et al., 2016). In 2018, the share of energy from renewable sources in gross final energy

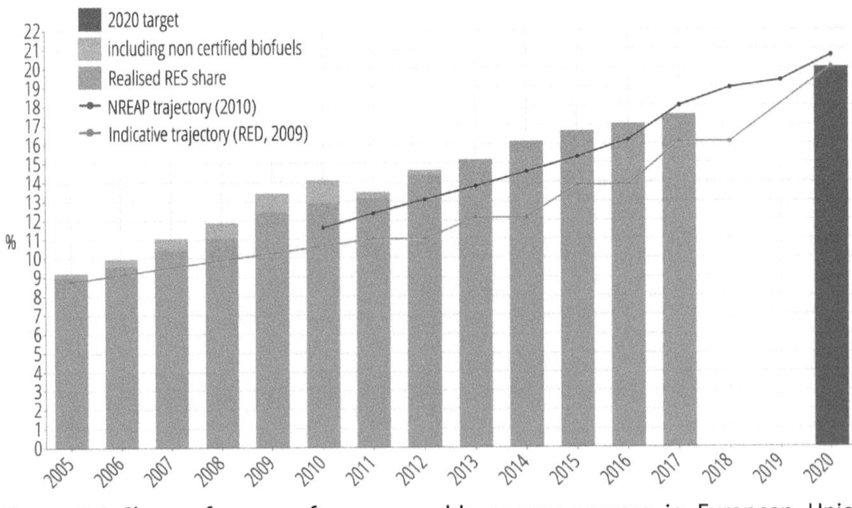

Figure 2.1 Share of energy from renewable energy sources in European Union (Zampori et al., 2016).

consumption reached 18.0% in the EU, increased from 17.5% in 2017 and more than double the share in 2004 (8.5%), the first year for which the data are available. The increase in the share of renewables is essential to reach the EU climate and energy goals. The EU's target is to reach 20% of its energy from renewable sources by 2020 and at least 32% by 2030. Especially for RES implementation on the building upgrade, the proportion of renewable energy production in the building stock, expressed as a percentage, varies from 25% (Cyprus) to 56% (Denmark) and 60% (Germany) (Zampori et al., 2016). The use of renewables for heating and cooling of buildings grew on average by 0.7% annually between 2005 and 2016, mainly by means of solid biomass and biogas, followed by solar thermal systems and, of course, by heat pumps. Finally, having a glance at the GHG emissions by sector, it is noticed that the building sector accounted for 36% of the GHG emissions in 2015 (Giama et al., 2020).

According to the Eurostat data released in January 2018, solar PV electricity output in the EU reached 123 TWh, which is 8.3% more than in 2017. Thus in 2018, solar PV accounted for 3.8% of the EU's gross electricity output (3.4% in 2017). In some countries, like Germany, Italy, and Greece, the solar power share is already more than 7% (7.1% in Germany, 7.8% in Italy, and 7.1% in Greece) (EurObserv, 2019). Focusing on PVs technology and the updated statistics, at least 103.2 GW of new PV capacity was installed across the world, which is a similar installation level to that of 2017. This new capacity raises global PV capacity to 512.3 GW, while in 1998, the global PV capacity was plateauing at just over 0.2 GW (234 MW). Therefore it is obvious that a significant progress has been made in all these years, and of course, there is an optimistic attitude for upcoming growth in the coming years (EurObserv, 2019).

In this chapter, the results of previous works are summarized, reviewed, and discussed. The environmental impact related to PV systems life cycle thinking is examined. The methodology followed in the analyzed case studies is the life cycle analysis (LCA), which offers a holistic approach to the environmental evaluation of systems.

2.2 PV system description

PV technology generates electricity from solar energy. Therefore, during the use phase, it offers clean energy with low GHG emissions. Nevertheless, it is important to evaluate the environmental impact of the

PV system in an LCA approach because of the energy-intensive manufacturing processes. Apart from the manufacturing processes related to significant GHG emissions, other processes are also examined within the LCA methodology, such as material transportation, PV system installation and retrofitting, and system disposal or recycling in terms of circularity.

PV systems can be described mainly by the material used in cells and secondly by the different types of applications and installations. Taking into account the various installation case studies, PV ground plants, PV roof systems, and facades were presented (review papers analysis Table 2.1). Common applications of PV technologies were examined in the building sector for reducing energy consumption and introducing clean energy to building stock.

The term building integrated photovoltaics (BIPVs) is used to describe PV elements, which can be used as materials and components to the building envelope. Additionally, they can be employed as building's roofs, giving energy benefit to the building upgrade and also architectural advantage promoting RES implementation based on EU energy and environmental policy (Jelle et al., 2012). The on-site electricity production can achieve significant energy savings and consequently reduce operational costs. This applies especially if the installation of the PV system does not need any extra construction structure; for instance, in facades, sometimes brackets and rails are needed depending on each system.

All these extra structural components increase the investment cost. Thus, if the initial cost is kept at a reasonable level, the benefits from the operational cost reduction are essential. Considering BIPVs, there is an extra categorization for PV technologies mainly for options that have been examined in most case studies, so the categorization refers to PV cells on slanted roofs, flat roofs, facades, and as shading systems. Further distinction for slanted roof and façade systems is further distinguished depending on the type of installation (building integrated using frameless laminate, or mounted using framed panel).

Apart from the different installation options, a critical categorization for PVs has to do with the material used in PV cells. Silicon is the most popular material used in PV modules. In the market, different types of silicon-based PV panels exist. The ones that are implemented in most PV systems is the crystalline silicon (c-Si) technology, which could be found in the form of single-crystalline silicon (sc-Si) (mono-crystalline is also used as a term for single-crystalline Si) or multicrystalline silicon (mc-Si)

Table 2.1 Overview/Scope—summary table of previous LCA studies.

References	Tool	LCI	LCIA	PV type	PV system	Boundaries	Fl. Chart	Fun. Unit	Country
Milousi et al. (2019)	SimaPro, EcoInvent, ReCiPe	Yes	Yes	• MCSi • PCSi • CIS • a-Si	• Roof-mounted	Cradle-to-grave	No	3 kWp	GR
Jungbluth (2005)	EcoInvent, Eco-Indicator	Yes	Yes	• MCSi • PCSi	• Roof-mounted • Façade	Cradle-to-gate	Yes	3 kWp	CH
Tripanagnostopoulos et al. (2005)	SimaPro, Eco-Indicator 95	No	No	• PCSi	• Roof mounted	Cradle-to-gate	No	3 kWp	GR
Carnevale et al. (2014)	ReCiPe	Yes	Yes	• MCSi • PCSi • CdTe • CIS	• Roof mounted	Cradle-to-grave	No	1 m^2	IT
Jungbluth et al. (2008)	EcoInvent, Eco-Indicator	No	No	• MCSi • PCSi • ribbon-Si • a-Si • CIS • CdTe	• Roof-mounted • Façade	Cradle-to-gate	No	3 kWp	CH
Frischknecht et al. (2015)	SimaPro	Yes	Yes	• MCSi • CdTe	• Roof-mounted	Cradle-to-grave	No	3 kWp	EU
Beylot et al. (2014)	IMPACT 2002 +	Yes	Yes	• PCSi	• Ground-mounted	Cradle-to-grave	Yes	5 MWp	FR
Pehnt (2006)	ReCiPe	Yes	Yes	• PCSi	• Roof-mounted • Ground-mounted	Cradle-to-grave	No	3 kW	DE

(Continued)

Table 2.1 (Continued)

References	Tool	LCI	LCIA	PV type	PV system	Boundaries	Fl. Chart	Fun. Unit	Country
Kunwor (2015)	GaBi	No	No	• c-Si • a-Si • CdTe • Ci(G)s	• —	Cradle-to-gate	Yes	5 m^2	FI
Sumper et al., 2011	SimaPro	No	Yes	• PCSi	• Roof-mounted	Cradle-to-gate	No	200 kWp	ES
Oliver and Jackson (2001)	ETH Zurich and Paul Scherrer Institute Database	No	No	• PCSi	• Façade • Ground-mounted	—	No	2–5 MWp	EU
Deciders et al. (2013)	Eco-Indicator 99	Yes	Yes	• MCSi	• Tracking system	Cradle-to-gate	No	1 MWp	IT
Pacca et al. (2007)	SimaPro	No	No	• PCSi • a-Si	• Roof-mounted	Cradle-to-gate	No	33 kWp	US
Stoppato (2008)	Boustead	Yes	Yes	• PCSi	• Roof-mounted	Cradle-to-gate	Yes	0.65 m^2	EU, AU, and US
Perez et al. (2012)	SimaPro	No	No	• MCSi	• Facade-integrated	—	No	153.5 m^2	US
Jungbluth et al. (2005)	Eco-Indicator 99	Yes	Yes	• MCSi • PCSi	• Roof-mounted • Façade	Cradle-to-gate	Yes	3 kWp	CH
Desideri et al. (2012)	SimaPro, Eco-Indicator 99	Yes	Yes	• PCSi	• Ground-mounted	Cradle-to-grave	Yes	1 kWp	IT
Bayod-Rújula et al. (2011)	SimaPro, Eco-Indicator 99	Yes	No	• PCSi	• Tracking system	Cradle-to-grave	Yes	26.46 kWp	ES
Menoufi et al. (2013)	Eco-Indicator 99 EPS 2000	Yes	Yes	• MCSi	• Façade Integrated	Cradle-to-gate	No	—	ES
Zhong et al. (2011)	Eco-Indicator 99	Yes	Yes	• PCSi	• Roof-mounted	Cradle-to-grave	Yes	—	NL

Graebig et al. (2010)	Eco-Indicator 99	Yes	Yes	• PCSi	• Ground-mounted	Cradle-to-gate	Yes	3 kWp	DE
Reich et al. (2011)	SimaPro	No	Yes	• c-Si	• —	Cradle-to-gate	No	—	NL
Mohr et al. (2013)	ReCiPe, SimaPro	No	Yes	• a-Si • nc-Si	• Roof-mounted	Cradle-to-gate	Yes	3 kWp	NL
Mason et al. (2006)	—	Yes	No	• PCSi	• Ground-mounted	Cradle-to-grave	No	1 MWp	US
Kato et al., 2001	—	No	No	• PCSi • a-Si • CdS/CdTe	• Roof-mounted	Cradle-to-gate	No	30 m^2	JP
Ito et al. (2010)	NEDO report	No	No	• MCSi • PCSi • a-Si • mc-Si • CIS • CdTe	• Ground-mounted	Cradle-to-gate	No	25 MWp	CN, South MO
Ito et al. (2008)	NEDO report	No	No	• PCSi • PCSi-he • a-Si • CdTe • CIS	• Ground-mounted	Cradle-to-gate	No	100 MW	CN, South MO
Held and Ilg (2011)	Gabi	No	Yes	• CdTe	• Ground-mounted	Cradle-to-grave	Yes	200 MWp	DE
Raugei et al. (2007)	SUMMA	Yes	Yes	• CdTe • CIS	• Roof-mounted	Cradle-to-gate	Yes	1 m^2	EU
Fthenakis and Kim (2005)	EcoInvent	No	No	• CdTe	• Ground-mounted	Cradle-to-gate	No	25 MWp	US
Alsema and de Wild-Scholten (2005b)	EcoInvent	No	Yes	• MCSi • PCSi • ribbon-Si	• Roof-mounted	Cradle-to-gate	No	1.25 m^2	South EU
Alsema and de Wild-Scholten (2005a)	SimaPro	No	No	• MCSi • PCSi • ribbon-Si	• Roof-mounted	Cradle-to-gate	No	1 kWp	South and Middle EU

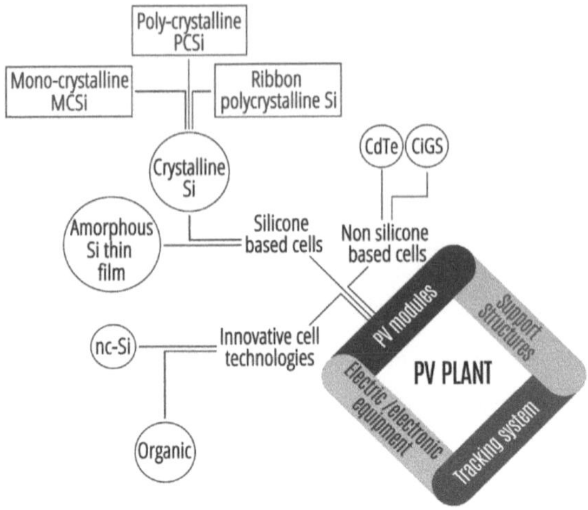

Figure 2.2 State-of-the-art research of photovoltaic technology.

form (poly-crystalline Si (PCSi) is also used as a term for multicrystalline PV panels). Moreover, another type of silicon-based PV cell is the amorphous Si (a-Si) thin-film panel and micro morph silicon (Fig. 2.2).

The different materials used in PV cells are related to differences concerning efficiency and cost-effectiveness. For instance, mono-crystalline silicon (MCSi) cells are tested to have more efficiency and highest initial cost compared to mc-Si cells, which have lower technical efficiency but reduced cost. Moreover, they are also examined to have differences in color, which has to do with manufacturing processes. For example, the poly-crystalline cells have a blue color due to the small crystals of the material used compared with the a-Si thin films that have a brownish or reddish-brown color.

Apart from the silicon-based PV panels, there is a second type of PV system based on thin-film PV technologies. These thin-film PV technologies—nonsilicone based—can further be divided into cadmium-telluride (CdTe), copper-indium-selenide (CIS), or copper-indium-gallium-diselenide (CIGS) cells. Second-generation PV systems are called "thin-film" systems because the semiconducting materials used to produce cells are only a few micrometers thick.

Apart from the material used in PV cells, some more parameters can affect the environmental impacts and deal with the supportive systems of a PV plant. For instance, the electric and electronic equipment is strongly related to the PV systems' efficiency as well as the environmental impact due to its relation to steel, copper, and plastic use. Moreover, the support structures used in roof and facades installations are subsystems related to materials use (plastic, aluminum, galvanized iron, copper) and energy consumption used in production processes of reflectors, heat exchangers, storage tank tubes and pipes, connected to the PV plant.

2.3 The methodology: life cycle analysis

The life cycle assessment (LCA) methodology evaluates and quantifies the environmental impacts for every stage of a product's life. The ISO 14040 and 14044 standards provide general guidance to implement an LCA.

LCA methodology aims at assessing the potential environmental impacts of a product or a service during its whole life cycle. According to ISO 14040, an LCA study shall be divided into the following main steps (Zampori et al., 2016):

- Goal and scope definition: in this stage, the functional unit (FU) and the system boundaries are determined.
- Life cycle inventory (LCI): the initial system is separated into different subsystems and energy, as well as the materials input, are quantified and listed.
- Life cycle impact assessment (LCIA): all the environmental impacts are evaluated for each flow from the LCI; a specific characterization factor determines its impact in the studied impact category. A specific score is finally obtained for each impact category studied. Normalization can also be used in this stage of analysis.
- Interpretation of the results: discussion of the results, conclusions, sensitivity analysis, improvement suggestions, and uncertainty.

In this chapter, the results of previous works are summarized, reviewed, and discussed. Assumptions relative to the LCA, for example, FU, system boundaries, or impact methodology, are also examined and presented targeting to give a detailed analysis of PVs' environmental impact. LCA is a methodology that evaluates and quantifies the environmental impacts, compares the environmental performance of different systems and technologies, points out processes with high environmental impact, and introduces measures in order to improve the system's environmental performance.

2.4 Inventory analysis

An inventory analysis in the LCA databases includes input data for:

- Energy consumption
- Materials, auxiliary chemicals, etc.

- Transport of materials
- Waste treatment
- Dismantling of all components

The lifetime of the system studied based on the literature review is 25–30 years, and the FU used in the systems studied is the kilowatt peak rating or the 1 m^2 of PV panel produced or installed.

Regarding the LCA methodology, the "cradle-to-grave" approach is followed, evaluating as well the GHG emissions related to the disposal phase. Nevertheless, in many studies, the "cradle-to-gate" approach is analyzed, targeting a more simplified approach and focusing mainly on the production processes of the system, which proved to be the more responsible for the increased environmental impacts due to significant energy consumption. Based on data analysis (Table 2.1), the SimaPro and the Eco-Indicator95, or the updated database of Eco-Indicator95, the Eco-Indicator99, are used in the majority of the case studies. Specifically, SimaPro and Eco-Indicator are used by about 30% more compared to the Ecoinvent database, which is used at about 16%, or Gabi, which is employed only at about 6%. The review results (Table 2.1) showed that in most cases, installing PV technologies in roofs and facades is preferable compared with, for example, the ground-mounted implementation. The BIPVs term is used for the PV technologies implementation in buildings as an energy conservation measure towards building upgrade, energy reduction, and CO_2 emissions decrease. When looking at the indicators estimated from the LCA implementation, the GHG emissions indicator is the one calculated in most studies as it is strongly related to energy consumption. The CO_2 emissions range depends on different PV technologies and installation from about 20 g/kWh up to 100 g/kWh (Table 2.2).

Over the last decades and as a result of the review analysis (Table 2.1), there have been a significant number of studies considering PV technologies. Especially in Europe, there is a sincere concern about evaluating and implementing RES technologies in compliance with the EU energy and environmental policy. Most of the LCA studies focus mainly on mono-crystalline and PCSi panels and from nonsilicone-based panels, CdTe, CIS, or CIGS cells is focused upon. The review analysis (Tables 2.1 and 2.2) also showed increased concern in the cradle-to-gate approach, mainly because the construction of PV cells is associated with high energy demand, which increases CO_2 emissions as well as GHG emissions.

2.5 Impact assessment

As discussed in the presentation of the technology of PVs section (Section 2.1), different types of PV modules require different processing. This significantly affects the environmental assessment of different panels, whereas apparently, the higher the degree of processing and chemicals used, the higher the environmental impact. With the lower processing degree, PV panels are the s-Di panels, followed by different thin-film technologies (CIS, CdTe). Poly-crystalline and mono-crystalline panels require higher levels of processing and quantities of chemicals. To this end, it is no surprise that the existing studies in the literature reveal that their impact on the environment is higher by definition. As a rule of thumb, the lower the efficiency of a panel, the lower its environmental impacts. In fact, LCA studies are hard to compare due to the different systems, boundaries, and FUs used. However, general conclusions may be extracted, revealing the fields which should be improved towards minimizing the environmental impact of the PV industry.

All PV modules' life cycle stages, from production to installation and treatment at the end of lifetime, result in environmental impacts. These can be calculated, separated into categories, and analyzed through LCIA methods. The environmental impacts evaluated from the different supportive tools and databases (Ecoinvent, Eco-Indicator, CML, etc.) are studied in most cases: global warming, acidification, eutrophication, photochemical oxidation, ozone layer depletion, human toxicity, freshwater aquatic ecotoxicity, and energy demand.

From the scientific literature, it may be concluded that the most energy-intensive and environmentally burdensome step among all stages and components of the whole PV chain is the production and operation phase of both thin-film and crystalline PV panels. As referred to Sumper et al. (2011), the production of the PV modules consumed more than 84% of the total primary energy consumption of the whole PV system. Compared to other phases, the effects caused by installation, maintenance, transports, and even disposal have shown to be unrelated, according to Carnevale et al. (2014). The energy consumption of the recycling phase is as high as for production and operating steps. However, the emissions and energy payback period associated with material recycling are also high, especially for the Si-based modules.

In Tables 2.3 and 2.4, the LCIA of Milousi et al. (2019) and Pehnt (2006), respectively, implemented with ReCiPe are presented. In both

Table 2.2 Overview of embodied carbon/embodied energy and EPBT.

References	CO_2 (g/kWh)	EPBT (years)
Milousi et al. (2019)	1. 43.5 g/kWh 2. 39.5 g/kWh 3. 44.3 g/kWh 4. 52.4 g/kWh	1. 10 years 2. 9.3 years 3. 9.6 years 4. 8.8 years
Jungbluth (2005)	• 39–110 g/kWh	• 3–6 years
Tripanagnostopoulos et al. (2005)	—	—
Carnevale et al. (2014)	—	1. 2.86 years 2. 2.33 years 3. 1.07 years 4. 0.97 years
Jungbluth et al. (2008)	• 40 g/kWh	—
Frischknecht et al. (2015)	1. 80 g/kWh 2. 30 g/kWh	1. 1.2–2.4 years 2. 0.7–1.1 years
Beylot et al. (2014)	• 37.5–53.5 g/kWh	—
Pehnt (2006)	• 100 g/kWh	—
Kunwor (2015)	• 54.9 g/kWh	—
Sumper et al. (2011)	—	• 3.5–5 years
Oliver and Jackson (2001)	—	1. 5.5 years 2. 8 years
Desideri et al. (2013)	• 44.7 g/kWh	• 5.5 years
Pacca et al. (2007)	• 34.3–72.4 g/kWh	• 3.2–7.5 years
Stoppato (2008)	• 4–840 g/kWh	• 3.4–6.5 years
Perez et al. (2012)	• 60.5 g/kWh	• 3.8 years
Jungbluth et al. (2005)	—	• 3–6 years
Desideri et al. (2012)	• 88.74 g/kWh	• 4.17 years
Bayod-Rújula et al. (2011)	—	• 1.45–1.5 years
Menoufi et al. (2013)	—	—
Zhong et al. (2011)	—	—
Graebig et al. (2010)	• 63 g/kWh	—
Reich et al. (2011)	Direct CO_2 emissions << indirect	—
Mohr et al. (2013)	—	• 2.3 years
Mason et al. (2006)	—	• 0.21 year
Kato et al. (2001)	• 13.4–19.7 g/kWh • 8.1–15.8 g/kWh • 8.9–14 g/kWh	• 1.6–2.4 years • 1.1–2.1 years • 1.1–1.7 years
Ito et al. (2010)	• 43–54 g/kWh	• 1.8–2.5 years
Ito et al. (2008)	• 12.1 g/kWh • 9.4 g/kWh • 15.6 g/kWh • 12.8 g/kWh • 10.5 g/kWh	• 1.9 years • 1.5 years • 2.5 years • 1.9 years • 1.6 years
Held and Ilg (2011)	• 19–30 g/kWh	• 0.7–1 1 years

(*Continued*)

Table 2.2 (Continued)

References	CO_2 (g/kWh)	EPBT (years)
Raugei et al. (2007)	• 17–48 g/kWh • 70–95 g/kWh	• 0.5–1.5 years • 1.9–2.8 years
Fthenakis and Kim (2005)	• 18–24 g/kWh	• 0.75–1.2 years
Alsema and de Wild-Scholten (2005b)	• 25–40 g/kWh	• 1.5–2.5 years
Alsema and de Wild (2005a)	• 30–45 g/kWh	• 1.7–2.7 years • 2.8–4.6 years

studies, a roof-mounted c-Si system of 3 kWp is analyzed. The boundaries of both studies are from cradle-to-grave. It is concluded that the wafer thickness has an impact on global warming potential (GWP). The impact assessment implemented in Carnevale et al. (2014) and Raugei et al. (2007) cradle-to-grave and cradle-to-gate boundaries, respectively, revealed that CIS and CdTe have less environmental impact compared to crystalline panels. Both studies were implemented for roof-mounted thin-film panels of 1 m^2 area. Raugei et al. (2007) concluded that CIS and CdTe have a less environmental impact than Si panels because fewer quantities of chemical compounds are used for their production, which are responsible for up to 20% of the overall impact indicators. Nevertheless, CIS and CdTe modules need more area for installation, mounting structure, and cabling per kilowatt hour due to their lower specific efficiencies. Values of current thin-film technology are quantitatively comparable with poly-crystalline technology, as opposed to mono-crystalline technology. Pehnt (2006) stated that a 50% reduction of the environmental impacts could be achieved by improving materials and energy supply.

As far as the energy payback period is concerned, the scientific literature reveals that the location of installation of the PV system plays a crucial role. Specifically, the higher the solar irradiation, the lower the energy payback period. Countries of the European south may have energy payback periods as low as 2 years. In contrast, the corresponding period for countries of northern Europe may be as high as 5 years (Carnevale et al., 2014). Certainly, the type of PV panel affects as well the time required for depreciation of the energy required for producing the panel (Tables 2.5 and 2.6).

Based on Jungbluth et al. (2008) and the different PV technologies studied, the LCIAresults showed that the environmental impacts of respiratory effects and fossil fuels depletion are the ones with the highest impact.

Table 2.3 LCIA—potential in impact categories according to ReCiPe tool.

Ref. - tool	PV type	Global warming (kg) CO_2-eq/kWh	Stratospheric ozone depletion kg $CFC11-eq/kWh$	Ionizing radiation kBq $Co\text{-}60-eq/kWh$	Ozone formation, human health kg NO_x-eq/kWh	Fine particular matter formation kg $PM_{2.5}-eq/kWh$	Ozone formation, terrestrial ecosystems kg NO_x-eq/kWh	Terrestrial acidification kg SO_2-eq/kWh	Freshwater eutrophication kg $P-eq/kWh$	Terrestrial ecotoxicity kg1,4 − DCB − eq / kWh	Freshwater ecotoxicity kg1,4 − DCB − eq / kWh	Marine ecotoxicity kg1,4 − DBC − eq / kWh	Human carcinogenic toxicity kg1,4 − DBC − eq / kWh	Human noncarcinogenic toxicity kg1,4 − DBC − eq / kWh	Land occupation m^2a crop − eq/kWh	Mineral resource scarcity kg $Cu-eq/kWh$	Fossil resource scarcity kg oil − eq / kWh	Water consumption m^3/kWh
Milousi et al. (2019)	■ MCSi	5.24×10^{-2}	2.45×10^{-8}	4.45×10^{-3}	1.20×10^{-4}	1.23×10^{-4}	1.25×10^{-4}	2.47×10^{-4}	4.07×10^{-5}	1.13	1.17×10^{-2}	1.54×10^{-2}	4.33×10^{-3}	1.64×10^{-1}	1.23×10^{-3}	5.42×10^{-4}	1.27×10^{-2}	1.17×10^{-3}
	■ PCSi	4.43×10^{-2}	2.06×10^{-8}	4.08×10^{-3}	1.05×10^{-4}	1.04×10^{-4}	1.10×10^{-4}	2.21×10^{-4}	3.78×10^{-5}	1.17	1.16×10^{-2}	1.53×10^{-2}	4.17×10^{-3}	1.63×10^{-1}	1.23×10^{-3}	5.54×10^{-4}	1.08×10^{-2}	1.35×10^{-3}
	■ a-Si	4.35×10^{-2}	1.70×10^{-8}	3.95×10^{-3}	9.83×10^{-5}	1.09×10^{-4}	1.01×10^{-4}	2.25×10^{-4}	3.55×10^{-5}	4.69×10^{-1}	1.11×10^{-2}	1.43×10^{-2}	6.50×10^{-3}	1.46×10^{-1}	1.13×10^{-3}	6.60×10^{-4}	1.04×10^{-2}	4.51×10^{-4}
	■ CIS	3.95×10^{-2}	1.75×10^{-8}	3.96×10^{-3}	9.09×10^{-5}	9.39×10^{-5}	9.26×10^{-5}	2.07×10^{-4}	4.62×10^{-5}	4.62×10^{-1}	1.30×10^{-2}	1.69×10^{-2}	4.19×10^{-3}	2.00×10^{-1}	9.60×10^{-4}	8.21×10^{-4}	9.40×10^{-3}	3.22×10^{-4}

Table 2.4 LCIA—potential in impact categories according to ReCiPe tool.

Ref. - Tool	PV Type	Global warming kg CO_2-eq / kWh	Ozone formation, human health kg NO_x-eq / kWh	Fine particular matter formation kg $PM_{2,5}-eq$ / kWh	Ozone formation, terrestrial ecosystems kg NO_x-eq / kWh	Terrestrial acidificationkg SO_2-eq / kWh	Freshwater eutrophication kg $P-eq$ / kWh	Mineral resource scarcity kg Bauxite − eq / kWh	Mineral resource scarcity kg Iron Ore − eq / kWh	Fossil resource scarcity kg oil − eq / kWh
Pehnt (2006)	■ PCSi	1.04×10^{-1}	1.90×10^{-6}	2.88×10^{-4}	3.40×10^{-4}	5.28×10^{-4}	4.4×10^{-5}	1.2×10^{-3}	3.3×10^{-3}	1.5 MJ (CED)

Table 2.5 LCIA—potential in impact categories implemented with Eco-Indicator and SUMMA scheme.

Ref. - Tool	PV Type	Global warming kg CO_2	Stratospheric ozone depletion kg CFC11	Terrestrial acidification kg SO_2	Eutrophication kg PO_4	Mineral extraction kg Pb	Carcinogens kg B(a)P	Summer smog kg C_2H_4	Winter smog kg SPM	Energy resources MJ LHV	Solid waste kg	Freshwater ecotoxicity kg1,4 – DCB – eq / kWh
Carnevale et al. (2014)	■ MCSi											
	i.	$-2.23\times10^{+3}$	-1.79×10^{-4}	$-1.11\times10^{+1}$	-1.68×10	-1.72×10^{-2}	-9.07×10^{-5}	-3.52×10^{-1}	-8.05×10	$-3.79\times10^{+4}$	$2.02\times10^{+1}$	
	ii.	$-2.49\times10^{+3}$	-1.90×10^{-4}	$-1.17\times10^{+1}$	-2.61×10	$-3.71\times\times10^{-2}$	-2.50×10^{-4}	-3.72×10^{-1}	$-1.20\times10^{+1}$	$-4.24\times10^{+4}$	3.05×10	
	■ PCSi											
	i.	$-2.28\times10^{+3}$	-1.81×10^{-4}	$-1.12\times10^{+1}$	-1.84×10	-2.05×10^{-2}	-9.32×10^{-5}	-3.56×10^{-1}	-8.13×10	$-3.88\times10^{+4}$	$2.03\times10^{+1}$	
	ii.	$-2.46\times10^{+3}$	-1.90×10^{-4}	$-1.17\times10^{+1}$	-2.49×10	-3.47×10^{-2}	-2.49×10^{-4}	-3.72×10^{-1}	-8.51×10	$-4.21\times10^{+4}$	3.05×10	
	■ CdTe											
	i.	$-1.95\times10^{+3}$	-1.54×10^{-4}	-9.04×10	-1.69×10	-9.47×10^{-3}	-8.52×10^{-5}	-2.95×10^{-1}	-6.51×10	$-3.35\times\times10^{+4}$	$3.81\times10^{+1}$	
	ii.	$-1.98\times10^{+3}$	-1.56×10^{-4}	-9.19×10	-1.68×10	-8.89×10^{-3}	-1.28×10^{-4}	-2.98×10^{-1}	-6.67×10	$-3.38\times10^{+4}$	3.64×10	
	■ CIS											
	i.	$-2.03\times10^{+3}$	-1.60×10^{-4}	-9.50×10	-1.89×10	-2.01×10^{-2}	-8.27×10^{-5}	-3.08×10^{-1}	-6.86×10	$-3.48\times10^{+4}$	$3.30\times10^{+1}$	
	ii.	$-2.07\times10^{+3}$	-1.64×10^{-4}	-9.82×10	-1.94×10	-2.27×10^{-2}	-2.06×10^{-4}	-3.14×10^{-1}	-7.18×10	$-3.54\times10^{+4}$	2.91×10	
Raugei et al. (2007)	■ CdTe											
	i.	1.7×10		3.0×10^{-2}								0.5×10^{-1}
	ii.	4.8×10		8.0×10^{-2}								3.0×10^{-1}
	■ CIS											
	i.	$7.0\times10^{+1}$		1.2×10^{-1}								3.4×10^{-1}
	ii.	$9.5\times10^{+1}$		1.5×10^{-1}								6.3×10^{-1}

Table 2.6 LCIA—potential in impact categories according to Eco-Indicator database.

Ref. - Tool	PV Type	Global warming kg CO_2	Stratospheric ozone depletion kg CFC11	Terrestrial acidification kg SO_2	Mineral extraction kg Pb	Carcinogens kg B(a)P	Fossil fuels kg oil − eq / kWh	Respiratory effects	Land occupation m^2a crop − eq / kWh	Ecotoxicity kg1,4 − DBC − eq / kWh
Jungbluth et al. (2008)	■ MCSi	0.35×10^{-3}	0.05×10^{-3}	0.10×10^{-3}	0.40×10^{-3}	0.30×10^{-3}	2.00×10^{-3}	1.20×10^{-3}	0.10×10^{-3}	0.35×10^{-3}
	■ PCSi	0.35×10^{-3}	0.05×10^{-3}	0.08×10^{-3}	0.40×10^{-3}	0.25×10^{-3}	1.90×10^{-3}	1.10×10^{-3}	0.10×10^{-3}	0.37×10^{-3}
	■ r-Si	0.43×10^{-3}	0.05×10^{-3}	0.08×10^{-3}	0.40×10^{-3}	0.30×10^{-3}	1.65×10^{-3}	1.15×10^{-3}	0.10×10^{-3}	0.37×10^{-3}
	■ a-Si	0.40×10^{-3}	0.07×10^{-3}	0.10×10^{-3}	0.60×10^{-3}	0.40×10^{-3}	1.55×10^{-3}	1.30×10^{-3}	0.10×10^{-3}	0.40×10^{-3}
	■ CdTe	0.30×10^{-3}	0.05×10^{-3}	0.12×10^{-3}	0.75×10^{-3}	0.55×10^{-3}	1.60×10^{-3}	1.50×10^{-3}	0.10×10^{-3}	0.48×10^{-3}
	■ CIS	0.35×10^{-3}	0.02×10^{-3}	0.08×10^{-3}	0.45×10^{-3}	0.25×10^{-3}	1.30×10^{-3}	1.00×10^{-3}	0.10×10^{-3}	0.32×10^{-3}

Global warming, mineral extraction, carcinogens, and ecotoxicity are following while land occupation and stratospheric ozone depletion are not proved to be significant compared to other environmental impacts. The LCA was conducted with the Eco-Indicator database (Tables 2.7 and 2.8).

Moreover, Frischknecht et al. (2015) concluded that human carcinogenic toxicity and energy demand contribute more than photochemical oxidation, global warming, terrestrial acidification, and land use occupation to the total environmental footprint of MCSi in three different studied scenarios. To sum up, it is not easy to select a specific PV technology based only on environmental criteria. The general conclusions lead to high energy demand construction processes, which is also related to significant carbon emissions (Table 2.9).

According to Anastaselos et al. (2016), the LCA was conducted with the cradle-to-gate approach. The analysis included the construction process and the transportation of raw materials. The FU used in this case is the 1 m^2 of the panels produced. The different types of PV technologies analyzed are the MCSi-based PV panels and the PCSi-based PV panels. The MCSi has more significant environmental impacts compared to PCSi. This can be explained by the complex manufacturing process of MCSi for mono-Si wafer production.

2.6 Conclusions—further research

The depletion of natural resources, the high intensive energy consumption from manufacturing processes, the increased GHG emissions from industries and other sectors have led to increased awareness towards environmental issues. Moreover, there is a defined EU strategy towards energy-saving measures, clean energy use, RES introduction, and reduced CO_2 emissions. National legislation framework and EU funding programs are in total compliance with EU energy and environmental policy. Thus, the selection of the most environmentally friendly solutions is inevitable and a key parameter to the decision-making process. The revised EU target for 2030 demands at least 40% reduction in GHG emissions from 1990 levels, at least 32% share for renewable energy, and at least 32.5% improvement in energy efficiency. Within the framework in this chapter, the environmental impact related to PV systems based on the life cycle thinking approach has been examined. The PV technologies were analyzed in detail and the LCA results from previous studies were grouped, described, and presented.

Table 2.7 LCIA—potential in impact categories according to SimaPro tool.

Ref. - Tool	PV Type	Global warming (%)	Fine particular matter (%) formation	Terrestrial acidification (%)	Human carcinogenic (%) toxicity	Land occupation (%)	Cumulative energy demand (%), non-renewable	Photochemical oxidation (%)
Frischknecht et al. (2015)	■ MCSi							
	i.	65	67	68	85	67	66	68
	ii.	31	27	25	57	39	41	34
	iii.	17	16	13	45	29	27	22
	CdTe							
	i.	69	75	83	94	74	71	69
	ii.	44	44	41	67	58	49	45
	iii.	32	33	29	54	49	36	33

Table 2.8 LCIA—potential in impact categories according to SimaPro tool.

Ref. - Tool	PV Type	Global warming kg $CO_2 - eq$ / kWh	Ozone formation, human health (kg) $NO_x - eq$ / kWh	Fine particular matter formation (kg) $PM_{10} - eq$ / kWh	Ozone formation, terrestrial ecosystems kg $NO_x - eq$ / kWh	Terrestrial acidification kg $SO_2 - eq$ / kWh	Mineral resource scarcity kg $Pb - eq$ / kWh
Sumper et al. (2011)	■ PCSi	$2.67 \times 10^{+5}$	1802.12	57.72	1802.12	$2.89 \times 10^{+3}$	0.13

Table 2.9 LCIA—potential in impact categories according to SimaPro tool.

Ref. - Tool	PV Type	Global warming $tnCO_2-eq/m^2$	Stratospheric ozone depletion g CFC-11eq/m^2	Eutrophication kg PO4− eq/m^2	Photochemical Ozone formation, kg C2H4 eq/m^2 kg NO_x − eq / kWh	Terrestrial ecotoxicity kg1,4 − DCB − eq/m^2	Freshwater ecotoxicity kg1,4 − DCB − eq/m^2	Marine ecotoxicity kg1,4 − DBC − eq / m^2	Human toxicity kg1,4 − Db − eq / m^2	Energy Demand 10GJeq/m^2
Anastaselos et al. (2016)	■ MCSi	0.1925	0.0436	0.1136	0.0401	0.0086	0.0178	0.0324	0.0094	0.026
	■ PCSi	0.1533	0.0423	0.0923	0.0331	0.0065	0.0145	0.0246	0.0086	0.0212

Most of the LCA studies focused mainly on MCSi panels, PCSi panels, and from nonsilicone-based panels such as CdTe, CIS, or CIGS cells. Increased concern is shown in the majority of the studies in the cradle-to-gate approach. This is due to the PV elements construction related to high energy demand, resulting in an increase of CO_2 emissions as well as GHG emissions. As it is already mentioned in the impact assessment analysis, LCA studies are rigid to compare due to the different systems, boundaries, and FUs used. However, general conclusions may be extracted, revealing the fields that should be improved towards minimizing the environmental impact of the PV industry ().

All PV modules' life cycle stages, from production to installation and treatment at the end of lifetime, result in environmental impacts. The majority of the articles concerning LCA of PV systems and technologies focus mainly on the cradle-to-gate approach and CO_2 emissions, which are also related to energy use. Thus, the carbon indicator is calculated in most of the studies as well as the GWP. Other indicators presented and discussed in the reviews' analysis are the energy flows and the energy payback time.

Future research targets to improve the environmental impacts of the materials used in PV cells focusing on high technical efficiency energy systems and cost-effective solutions. Within this framework and in terms of circular economy, new generation PV types will be introduced, increasing recyclability and reuse perspectives in the construction process as well as energy-saving opportunities in the production process. Organic panels toward environmental impacts reduction are examined. Finally, further research is targeting flexible and lightweight materials, thin-film solar cells in order to facilitate the installation procedure and avoid heavy support structures for roofs and facades applications.

Nomenclature

LCI life cycle inventory
LCIA life cycle impact assessment
FU functional Unit
EPBT energy payback time
MCSi mono-crystalline Silicone
PCSi poly-crystalline silicone
PCSi-he poly-crystalline silicone—high efficiency
a-Si amorphous Silicone
c-Si crystalline Silicone
CIS copper Indium Selenium

Ci(G)s copper-indium-gallium-diselenide
CdTe cadmium-Telluride
nc-Si nano-crystalline Silicone
mc-Si micro-crystalline Silicone

References

Alsema, E., de Wild, M.J., 2005a. Environmental impact of crystalline silicon photovoltaic module production. MRS Online Proceedings Library Archive 895. Available from: https://doi.org/10.1557/PROC-0895-G03-05.

Alsema, E.A., de Wild-Scholten, M.J., 2005b. The real environmental impacts of crystalline silicon PV modules: an analysis based on up-to-date manufacturers data, in Pesented at the 20th European Photovoltaic Solar Energy Conference, Vol. 6. EDTEWEB, Netherlands, p. 10.

Anastaselos, D., Oxizidis, S., Manoudis, A., Papadopoulos, A.M., 2016. Environmental performance of energy systems of residential buildings: toward sustainable communities. Sustainable Cities and Society 20, 96–108.

Bayod-Rújula, Á.A., Lorente-Lafuente, A.M., Cirez-Oto, F., 2011. Environmental assessment of grid connected photovoltaic plants with 2-axis tracking vs fixed modules systems. Energy 36 (5), 3148–3158.

Beylot, A., Payet, J., Puech, C., Adra, N., Jacquin, P., Blanc, I., et al., 2014. Environmental impacts of large-scale grid-connected ground-mounted PV installations. Renewable Energy 61, 2–6.

Carnevale, E., Lombardi, L., Zanchi, L., 2014. Life cycle assessment of solar energy systems: comparison of photovoltaic and water thermal heater at domestic scale. Energy 77, 434–446.

Desideri, U., Proietti, S., Zepparelli, F., Sdringola, P., Bini, S., et al., 2012. Life cycle assessment of a ground-mounted 1778 kWp photovoltaic plant and comparison with traditional energy production systems. Applied Energy 97, 930–943.

Desideri, U., Zepparelli, F., Morettini, V., Garroni, E., et al., 2013. Comparative analysis of concentrating solar power and photovoltaic technologies: technical and environmental evaluations. Applied Energy 102, 765–784.

EurObserv, E.R., 2019. The State of Renewable Energies in Europe, 19th EurObserv'ER Report, Observ'ER (FR), TNO Energy Transition (NL), RENAC (DE), Frankfurt School of Finance and Management (DE), Fraunhofer ISI (DE) and Statistics Netherlands (NL) <http://www.eurobserv-er.org/> (accessed November 11, 2020).

Frischknecht, R., Itten, R., Wyss, F., Blanc, I., Heath, G.A., Raugei, M., et al., 2015. Life Cycle Assessment of Future Photovoltaic Electricity Production From Residential-Scale Systems Operated in Europe (No. NREL/TP-6A20-73849). National Renewable Energy Lab. (NREL), Golden, CO.

Fthenakis, V., Kim, H.C., 2005. Energy use and greenhouse gas emissions in the life cycle of CdTe photovoltaics. MRS Online Proceedings Library Archive 895.

Giama, E., Kyriaki, E., Papadopoulos, A.M., 2020. Energy policy and regulatory tools for sustainable buildings. IOP Conference Series: Earth and Environmental Science 410 (1), 012078. IOP Publishing.

Graebig, M., Bringezu, S., Fenner, R., 2010. Comparative analysis of environmental impacts of maize–biogas and photovoltaics on a land use basis. Solar Energy 84 (7), 1255–1263.

Held, M., Ilg, R., 2011. Update of environmental indicators and energy payback time of CdTe PV systems in Europe. Progress in Photovoltaics: Research and Applications 19 (5), 614–626.

Ito, M., Kato, K., Komoto, K., Kichimi, T., Kurokawa, K., et al., 2008. A comparative study on cost and life-cycle analysis for 100 MW very large-scale PV (VLS-PV) systems in deserts using m-Si, a-Si, CdTe, and CIS modules. Progress in Photovoltaics: Research and Applications 16 (1), 17−30.

Ito, M., Komoto, K., Kurokawa, K., 2010. Life-cycle analyses of very-large scale PV systems using six types of PV modules. Current Applied Physics 10 (2), S271−S273.

Jelle, B.P., Breivik, C., Røkenes, H.D., 2012. Building integrated photovoltaic products: a state-of-the-art review and future research opportunities. Solar Energy Materials and Solar Cells 100, 69−96.

Jungbluth, N., 2005. Life cycle assessment of crystalline photovoltaics in the Swiss ecoinvent database. Progress in Photovoltaics: Research and Applications 13 (5), 429−446.

Jungbluth, N., Bauer, C., Dones, R., Frischknecht, R., et al., 2005. Life cycle assessment for emerging technologies: case studies for photovoltaic and wind power (11 pp). The International Journal of Life Cycle Assessment 10 (1), 24−34.

Jungbluth, N., Tuchschmid, M., de Wild-Scholten, M., 2008. Life Cycle Assessment of Photovoltaics: Update of Ecoinvent Data V2. 0, ESU-services Ltd., 50.

Kato, K., Hibino, T., Komoto, K., Ihara, S., Yamamoto, S., Fujihara, H., et al., 2001. A life-cycle analysis on thin-film CdS/CdTe PV modules. Solar Energy Materials and Solar Cells 67 (1−4), 279−287.

Kunwor, A., 2015. Economic and Environmental Evaluation of Renewable Energy Systems (PhD thesis). Lappeenranta University of Technology.

Mason, J.E., Fthenakis, V.M., Hansen, T., Kim, H.C., et al., 2006. Energy payback and life-cycle CO_2 emissions of the BOS in an optimized 3• 5 MW PV installation. Progress in Photovoltaics: Research and Applications 14 (2), 179−190.

Menoufi, K., Chemisana, D., Rosell, J.I., 2013. Life cycle assessment of a building integrated concentrated photovoltaic scheme. Applied Energy 111, 505−514.

Milousi, M., Souliotis, M., Arampatzis, G., Papaefthimiou, S., et al., 2019. Evaluating the environmental performance of solar energy systems through a combined life cycle assessment and cost analysis. Sustainability 11 (9), 2539.

Mohr, N.J., Meijer, A., Huijbregts, M.A.J., Reijnders, L., et al., 2013. Environmental life cycle assessment of roof-integrated flexible amorphous silicon/nanocrystalline silicon solar cell laminate. Progress in Photovoltaics: Research and Applications 21 (4), 802−815.

Oliver, M., Jackson, T., 2001. Energy and economic evaluation of building-integrated photovoltaics. Energy 26 (4), 431−439.

Pacca, S., Sivaraman, D., Keoleian, G.A., 2007. Parameters affecting the life cycle performance of PV technologies and systems. Energy Policy 35 (6), 3316−3326.

Pehnt, M., 2006. Dynamic life cycle assessment (LCA) of renewable energy technologies. Renewable Energy 31 (1), 55−71.

Perez, M.J., , et al.,Fthenakis, V., Kim, H.C., Pereira, A.O. 2012. Façade−integrated photovoltaics: a life cycle and performance assessment case study. Progress in Photovoltaics: Research and Applications 20 (8), 975−990.

Raugei, M., Bargigli, S., Ulgiati, S., 2007. Life cycle assessment and energy payback time of advanced photovoltaic modules: CdTe and CIS compared to poly-Si. Energy 32 (8), 1310−1318.

Reich, N.H., Alsema, E.A., Van Sark, W.G.J.H.M., Turkenburg, W.C., Sinke, W.C., et al., 2011. Greenhouse gas emissions associated with photovoltaic electricity from crystalline silicon modules under various energy supply options. Progress in Photovoltaics: Research and Applications 19 (5), 603−613.

Stoppato, A., 2008. Life cycle assessment of photovoltaic electricity generation. Energy 33 (2), 224−232.

Sumper, A., Robledo-García, M., Villafáfila-Robles, R., Bergas-Jané, J., Andrés-Peiró, J., et al., 2011. Life-cycle assessment of a photovoltaic system in Catalonia (Spain). Renewable and Sustainable Energy Reviews 15 (8), 3888–3896.
Tripanagnostopoulos, Y., Souliotis, M., Battisti, R., Corrado, A., et al., 2005. Energy, cost and LCA results of PV and hybrid PV/T solar systems. Progress in Photovoltaics: Research and Applications 13 (3), 235–250.
Zampori, L., Saouter, E., Schau, E., Cristobal Garcia, J., Castellani, V., Sala, S., et al., 2016. Guide for Interpreting Life Cycle Assessment Result. Publications Office of the European Union, Luxembourg.
Zhong, Z.W., Song, B., Loh, P.E., 2011. LCAs of a poly-crystalline photovoltaic module and a wind turbine. Renewable Energy 36 (8), 2227–2237.

CHAPTER 3

Life cycle assessment review in solar thermal systems

Maria Milousi and Manolis Souliotis
Department of Chemical Engineering, University of Western Macedonia, Koila, Greece

Contents

3.1 Introduction

Solar energy is a clean, renewable energy source that can be used in several ways and a variety of applications, replacing energy with fossil fuels. Given that solar energy conversion into heat through solar thermal systems has a typical efficiency of between 40% and 60%, depending on the desired operating temperature, its use for direct heat production is the most appropriate energy option. Solar thermal technologies are widespread and contribute significantly to reducing household energy demand. They are essentially engaged in producing hot water for domestic use and can meet significant thermal needs. However, despite their advantages, these systems must modify their design priorities; otherwise, they will inevitably undermine their sustainable situation without effective end-of-life management.

An important factor in the development of solar thermal systems is the ability to harness the cheap and abundant energy of the sun using minimal surface collection. On the contrary, both their construction and their final

Environmental Assessment of Renewable Energy Conversion Technologies
DOI: https://doi.org/10.1016/B978-0-12-817111-0.00006-1

disposal are linked to environmental changes. This is due to the energy required for the extraction of raw materials, the assembly of the finished product, as well as the final disposal and/or recycling of the collector. It is, therefore, necessary to assess the indirect environmental impacts caused by the provisions throughout their life cycle (Milousi et al., 2019; Arnaoutakis et al., 2019).

The following review is divided into two sections: whether the solar thermal system is integrated into the building from the beginning of its construction or installed afterward. In this way, there are the building-integrated solar thermal (BIST) collectors and the building-added solar thermal systems.

3.2 Building-integrated solar thermal collectors

3.2.1 Flat plate solar thermal collectors

Lamnatou et al. (2014) conducted a life cycle assessment (LCA) of a patented flat plate BIST collector developed and experimentally tested at the University of Corsica in France. Three configurations (System 1, 2, 3) were examined: System 1 with collectors in series connection; tubes at different levels, System 2 with collectors in parallel connection; tubes at different levels and System 3 with collectors in series connection; tubes at the same level, while multiple scenarios were adopted, that is, recycling/no recycling. The results revealed for the systems, embodied energy (EE) of 3 GJ_{prim}/m^2 and embodied carbon (EC) of 0.16 $tCO_2equiv./m^2$, for the scenario "no recycling," whereas by means of recycling, the corresponding values decreased to approximately 0.4–0.5 GJ_{prim}/m^2 and 0.02–0.03 tCO_2eq/m^2. Under the same situation, the energy payback time (EPBT) of system 1 was approximately 1.5 years, while by utilizing system 2, this value decreased to around 0.5 years. Additionally, a remarkable reduction (28%–96%) in CO_2equiv. emissions of all systems was achieved by adopting configurations with double collector surface/output. Regarding the indicator of sustainability (IS), system 2 showed a value of 0.78, close to that of the "ideal" system 3 (IS = 1), and the authors claimed that system 2 could exist in practice (Fig. 3.1).

Lamnatou et al. (2015) provided an investigation about the environmental profile of a patented flat plate BIST active system, based on multiple life cycle impact assessment (LCIA) methodologies that were developed and experimentally tested at the University of Corsica in France. Three configurations were examined: System 1 was connected in

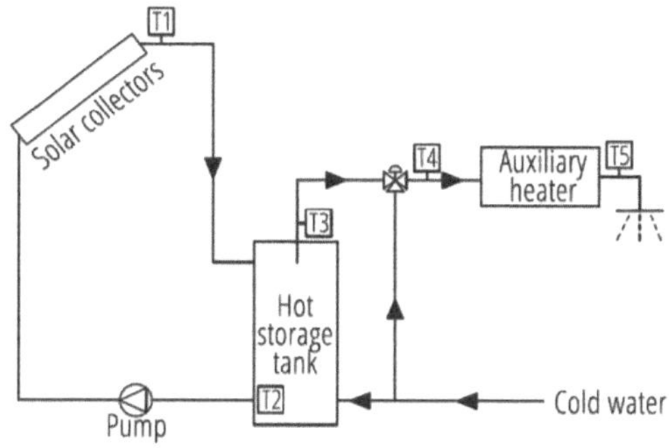

Figure 3.1 Schematic diagram of an solar hot water system. *Adopted from Hang, Y., Qu, M. and Zhao, F., 2012. Economic and environmental life cycle analysis of solar hot water systems in the United States, Energy and Buildings, 45, 181–188.*

series (for cold and hot water) at different levels, in System 2, the collectors were in parallel at different levels, and System 3 was connected in series with the tubes at the same level, while multiple approaches were examined that is, recycling/no recycling. By considering the full life cycle in terms of Eco-indicator 99 (EI99), the footprint values of all systems indicated that by utilizing System 2 instead of system 1, an impact reduction of 10%–13% was achieved (depending on the damage category), while by using System 3 instead of System 1 this reduction ranged from 12% to 17%. Moreover, by comparing all the studied cases, that is, full life cycle, Systems 1–3, and Scenarios no recycling versus recycling, it was demonstrated that recycling led to an impact reduction of 8%–19%. The results from the evaluation of the systems from the EE/EC point of view revealed that system 1 could considerably improve its environmental performance by utilizing collectors connected in parallel (System 2). In addition, the EPBT of System 1 decreased to less than 2 years by connecting the collectors in parallel and exhibited values around 0.5 years if recycling was also considered. The EC of the systems was around 0.16 tCO_2equiv./m^2 without recycling and 0.02–0.03 tCO_2equiv./m^2 with recycling. The authors concluded that System 2 had significantly improved efficiency and environmental performance. At the same time, a further reduction of the impact could be achieved by using recycling and/or systems with double absorber surface and double output (Fig. 3.2).

Lamnatou et al. (2018a) evaluated the environmental profile of a flat plate BIST system, based on multiple LCIA methods and two basic scenarios (in terms of heat storage/insulation): with and without phase change material (PCM) in France. The following results were reported in

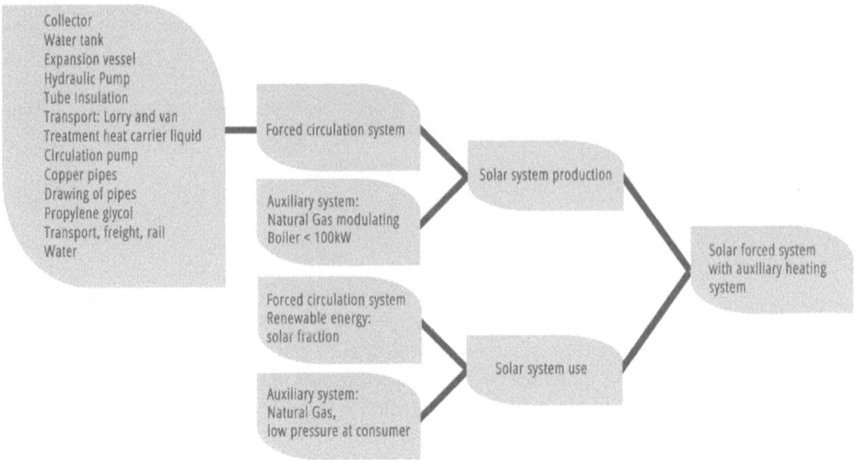

Figure 3.2 System boundary for a solar thermal system with forced circulation. *Adeopted from Alberti, J., Raigosa, J., Raugei, M., Assiego, R., Ribas-Tur, J., Garrido-Soriano, N., et al., 2019. Life cycle assessment of a solar thermal system in Spain, eco-design alternatives and derived climate change scenarios at Spanish and Chinese national levels. Sustainable Cities and Society, 47, 101467.*

the phase of material manufacturing. The ReCiPe endpoint values revealed the responsible materials for the highest impact: the tubes (copper) in the human health category, the aluminum parts, and the PCM component in ecosystems. In contrast, the PCM component presented considerably higher value in ecosystems than in the human health category. Moreover, based on midpoint categories of ReCiPe, aluminum, copper, and PCM were the three materials with the highest values in terms of the total impact of the whole system. Furthermore, according to USEtox findings among all the components/materials, the PCM component presented the highest values in both categories of human toxicity (noncancer and cancer) and ecotoxicity. Especially in the case of ecotoxicity, there was a remarkable difference between the PCM component and the other materials. Regarding ecological footprint (EF) results, among all the components/materials, the aluminum-based parts (absorber, casing, gutter) presented the highest impact (Fig. 3.3).

Lamnatou et al. (2018b) implemented an LCA of a flat plate BIST system, based on multiple LCIA methods with two configurations in terms of the collectors (with/without PCM) in France. The material manufacturing phase (collectors, additional components, replacements over lifespan), based on cumulative energy demand (CED) and global

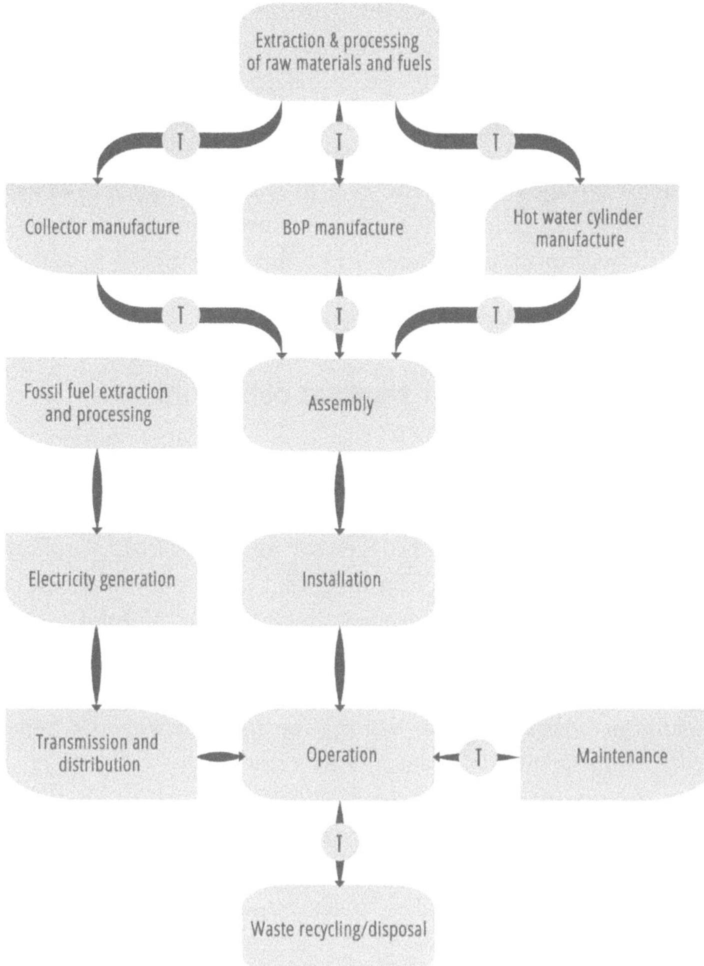

Figure 3.3 The life cycle of solar thermal systems. *Adopted from Greening, B. and Azapagic, A., 2014. Domestic solar thermal water heating: a sustainable option for the UK? Renewable Energy, 63,23–36.*

warming potential (GWP), revealed that the three materials with the highest impact were aluminum, steel, and PCM. The configuration without PCM exhibited values of 0.06 kg CO2equiv./kWh (based on GWP 100a) and 0.67 MJ_{prim}/kWh for the life cycle results and the scenario without recycling. The configuration with PCM presented 0.08 kg CO_2equiv./kWh (based on GWP 100a) and 0.74 MJ_{prim}/kWh. On the other hand, in scenarios with recycling, the findings demonstrated that recycling improved the environmental profile of the proposed BIST system considerably. Moreover, by taking into account the inputs for pumping/auxiliary heating, the EPBT was calculated to be 4.24 and 3.68 years for the configuration without and with PCM. Therefore, PCM offered an EPBT reduction of around 0.6 years, which occurred mainly due to the fact that PCM allowed the use of a lower flow rate. However, if the inputs for pumping/auxiliary heating were not considered, both configurations (with/without PCM) had almost the same EPBT (approximately 1.3 years) (Fig. 3.4).

3.2.2 Evacuated tube solar thermal collectors

Lamnatou et al. (2019) evaluated a vacuum tube BIST system in France through LCA, based on ReCiPe midpoint/endpoint and USEtox in relation to different electricity mixes with/without recycling scenarios. According to the scenario without recycling, the ReCiPe midpoint results indicated that when considering material manufacturing of all the collectors, the copper-based ones presented the highest impact among glass-, aluminum-, copper-, and steel-based components 15 of the 18 impact categories. On the other hand, regarding USEtox and when taking into account both the collectors and the additional elements of the system, the material with the highest total score for (a) human toxicity/cancer and ecotoxicity was copper and (b) human toxicity/noncancer was propylene glycol. Again with the scenario without recycling, ReCiPe payback time showed values of 18.14 and 4.03 years in the case of France's and Spain's electricity mix, respectively. On the contrary, through recycling of glass, metals, and plastics, the corresponding values were reduced to 15.48 and 3.44 years, respectively, resulting that recycling offered a remarkable reduction in ReCiPe payback time. A similar trend regarding the USEtox findings was observed with a 20%–95% decrease in the environmental impact, depending on the metal recycled type and the impact category (Fig. 3.5).

Lamnatou et al. (2016) investigated an LCA comparison of a vacuum tube/BIST system with a flat plate/BIST configuration. Both systems were gutter integrated and patented and they had been developed and tested at the University of Corsica in France. Multiple LCIA methodologies, environmental indicators, scenarios, and databases were adopted. Among them System I: flat plate collectors of parallel connection with tubes at different levels and system II: vacuum-tube collectors interconnected with copper tube. The findings revealed that system I exhibited EPBT values of 1.4 and 0.4 years

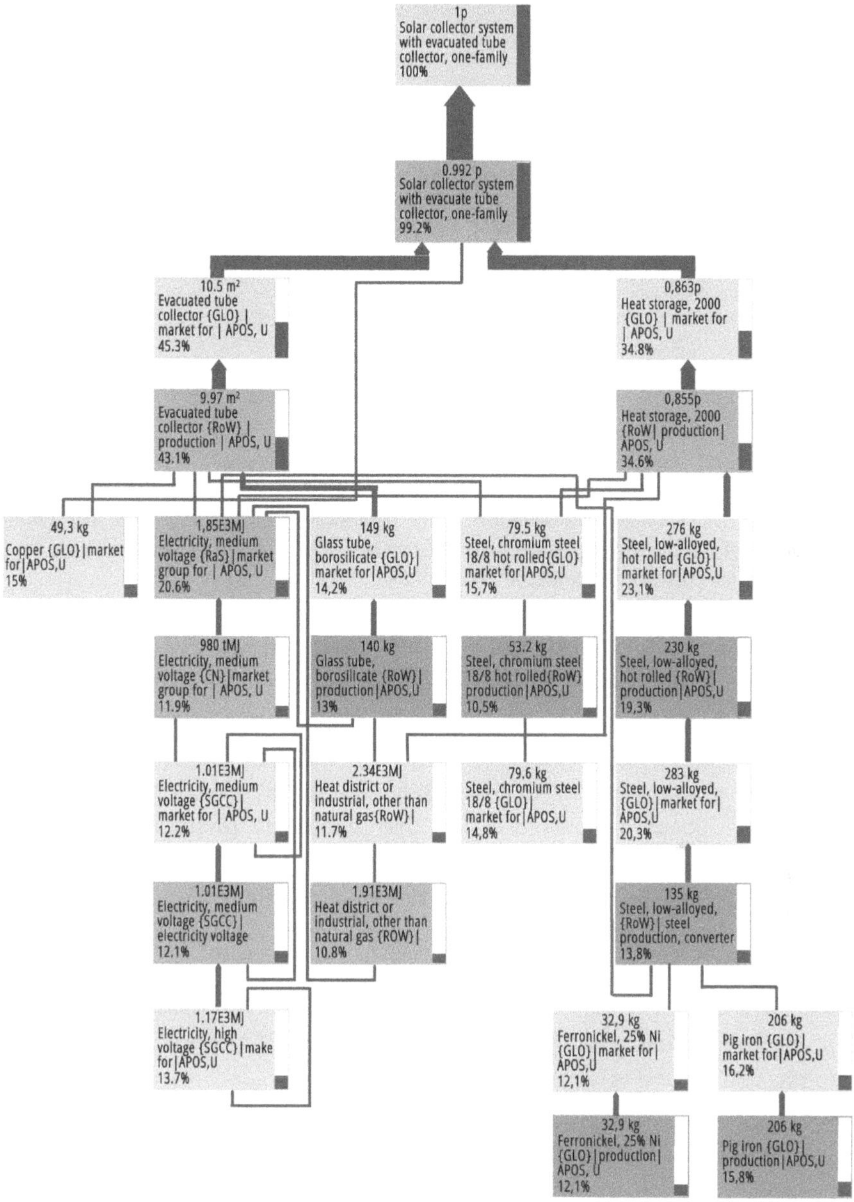

Figure 3.4 Process network for a vacuum tube solar collector. *Adopted from Milousi, M., Souliotis, M., Arampatzis, G. and Papaefthimiou, S., 2019. Evaluating the environmental performance of solar energy systems through a combined life cycle assessment and cost analysis. Sustainability, 11(9), 2539.*

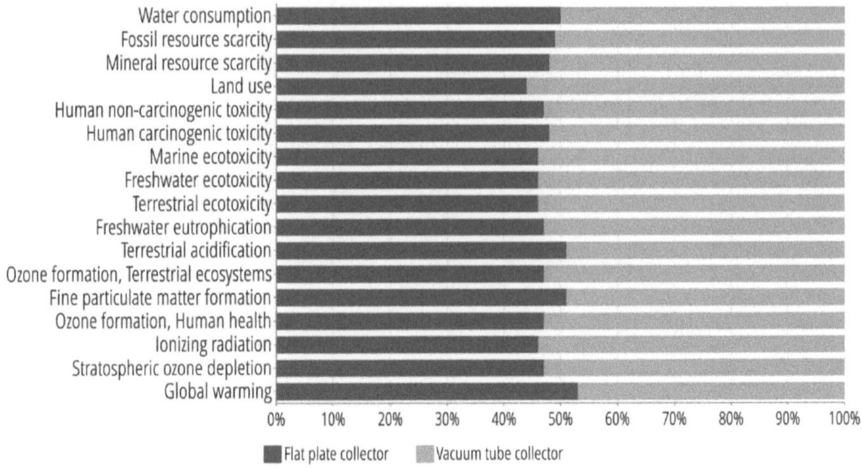

Figure 3.5 Comparative LCA results of a flat plate and vacuum tube system. *Adopted from Milousi, M., Souliotis, M., Arampatzis, G. and Papaefthimiou, S., 2019. Evaluating the environmental performance of solar energy systems through a combined life cycle assessment and cost analysis. Sustainability, 11(9), 2539.*

higher (for the cases without and with recycling, respectively) than system II. Adopting recycling for the evaluation of the EPBT led to a reduction of 1.3 and 0.3 years for System I and System II, respectively. According toenergy return on investment (scenario without recycling and 30-year lifespan), the calculations revealed values of 16.5 and 64.2 for System I and System II, respectively. These findings revealed that system II presented remarkably higher long-term viability in comparison to System I. With respect to greenhouse gas payback time (GPBT) (case without recycling), System I presented values of 1.4–13.7 years, while these values for System II were 0.5 to 4.1 years. On the other hand, by adopting recycling to evaluate the GPBT, there was a reduction of 0.4–1.8 years. Based on USEtox, EF, and France's electricity mix—as well as by having as reference the CO_2equiv. emissions of a domestic gas boiler—the calculations about the avoided impact during the use phase of the proposed systems demonstrated that considerably higher impact could be avoided by utilizing System II instead of System I.

3.3 Building-added solar thermal systems

3.3.1 Flat plate solar thermal collectors

Kalogirou (2009) studied the environmental and economic benefits of utilizing solar energy instead of conventional energy sources in Nicosia,

Cyprus. In particular, an LCA was conducted for a thermosiphon solar water heater with two types of backup: electricity and diesel. The results indicated that the total energy used for the manufacture and installation of the solar system is recouped in about 13 months. In contrast, the payback time concerning emissions produced from the EE required for the manufacture and installation of the systems varied from a few months to 3.2 years, depending on the fuel and the particular pollutant considered. Moreover, the payback time was estimated to be 2.7 and 4.5 years with electricity and diesel back up, respectively, and the corresponding life cycle savings were 2,240 € and 1,056 €.

Moore et al. (2017) implemented an LCA investigating the GWP and primary energy demand (PED) of five common domestic hot water systems installed in Australia. The results indicated that the typical and most commonly installed hot water system that is electric storage exhibited the highest carbon footprint. Three of the studied systems were based on conventional energy resources, that is, nonsolar. At the same time, the other two were solar thermosiphon systems with supplementary water heating provided by an electric immersion element and an inline gas instantaneous booster, respectively. The electric storage hot water system had the highest GWP and PED, followed by solar electric, gas storage, gas instantaneous, and then solar gas instantaneous system. The use stage of the life cycle accounted for 87%–99% of the GWP or PED in all systems. The impacts from the production of the hot water systems were relatively minor. As a result, the GWP and PED payback periods were less than 12 months for all solar systems compared to their nonsolar equivalent.

Koroneos and Nanaki (2012) carried out an LCA and economic analysis of a flat plate solar water heating system with electricity used as an additional auxiliary source, located in Thessaloniki, Greece. The environmental findings revealed that the significant contribution came from the storage tank, accounting for 58% of the total environmental impact, the solar collector accounted for 25%, whereas the aperture and component box had more negligible impacts, 10%, and 7%, respectively. The main environmental impact of the studied system was acidification, with a percentage reaching up to 54%. The second highest environmental effect was the winter smog potential, with a value reaching 25%. The contribution to the GWP due to CO_2 was also presented in a lower percentage of 12%. Moreover, the use of the solar water heating system exhibited savings equal to 4,280€ over its lifetime with a payback period equal to 5 years.

Comodi et al. (2014) performed an LCA for two domestic solar hot water systems (DSHWS), that is, a traditional model with glazed panels and a system with unglazed solar collectors for three different locations; Rome, Madrid, and Munich. The energy, CO_2 emissions, and economic payback time were calculated for each system, keeping as reference basis of comparison two competitive energy technologies: natural gas and electric boiler. The results indicated that for the traditional DSHWS, 93% of the impacts were due to panel production. The storage tank and the panel production resulted in more balanced impacts for the unglazed system: 54% and 44%, respectively. The EPBT ranged between 2 and 12 months, while GPBT ranged between 1 and 30 months. The economic payback time for the natural gas and the electrical boiler ranged between 8 and 13 years, and 3 and 4 years, respectively.

Piroozfar et al. (2016) reported findings from a comparative cradle-to-grave LCA of five DSHWS in the UK. Amongst them, a solar heater with electric backup appeared to be the least damaging alternative. Specifically, once the percentages had been normalized, the typical electric heater shower was the DSHWS with the highest environmental impact benchmarked at 100%, followed by the passage heater with gas at 98%, the gas boiler at 68%, the electric boiler at 57%, and finally the solar heater with electric back up at 49%. The results of a cradle-to-gate ranking for the studied DSHWS would be the exact opposite compared to that from a cradle-to-grave study. For example, the solar system would be the one with the most severe impacts, followed by the electric boiler at 73% of the solar system, then the gas boiler at 70%, the passage heater with gas at 56%, and the electric shower at 43%. The authors concluded that a cradle-to-gate study might lead to choosing the most damaging alternative despite the probably genuine aim of identifying the least damaging one. Therefore, adopting a cradle-to-grave perspective when conducting an LCA to inform about environmental burdens conclusions is vital.

Alberti et al. (2019) compared the environmental performance of a DSHWS used in conjunction with a traditional natural gas heating system and the natural gas heating system on its own in Barcelona, Spain. In addition, the authors explored the national GHG emissions reduction potential of broader use of DSHWS in China's and Spain's built environment. In this case, five displacement methods were suggested to show how the emissions reduction varied. The solar system presented an important improvement in the categories of global warming, ozone depletion, and formation of photochemical ozone. On the other hand, their fundamental weak points were in the acidification and eutrophication categories, in which impacts were shown to be higher compared to the conventional systems. In particular, the water tank, the collector, and the copper tubes of the secondary circuit were revealed to be the components with the largest environmental impact. Moreover, a potential shift to renewable technologies of 22.2% of the energy used in domestic hot water (DHW) would imply a reduction corresponding between 0.38% and 0.5% of the total CO_2equiv. emissions in Spain (data for the year 2015). By replacing electricity and natural gas with solar thermal technologies for DHW in different Chinese national scenarios, a reduction between 0.119% and 0.629% of the Chinese total CO_2equiv. emissions can be achieved.

Altun-Çiftçioğlu et al. (2016) focused on the LCA of a roll to roll (R2R) continuous solar selective surface production for high-efficiency solar thermal collectors in Turkey. For that reason, two different flat solar collectors were compared using LCA for the different production technology of the selective surfaces, that is, batch production or R2R. The 200,000 m^2 net selective surface production results exhibited a CED of 7.91 TJ and 64,900 kg of total CO_2equiv. emissions per year. Furthermore, using the R2R process for a complete flat solar collector showed a reduction of 1.2% and 0.8% in CED and GHG emissions, respectively. At the same time, the payback time of the selective surface production only would be 3–4 days. Moreover, choosing the new technology system decreased the overall total environmental impact by 1.3%. In particular, when only the selective surface production process was considered, the R2R continuous process decreased this burden by 52%.

Morsink-Georgali et al. (2017) conducted a comparative LCA and evaluated the environmental impacts of a manufacturing system for solar collectors by investigating four alternative scenarios in Cyprus. In these systems, the back cover and the frame were manufactured either with aluminum alloy (Scenario 1) or recycled aluminum (Scenario 2). In Scenario 3, the absorber was manufactured from copper instead of aluminum. In Scenario 4, the combination of Scenarios 2 and 3 was assumed. The results confirmed that the environmental footprint of a solar thermal panel could significantly be affected by the use of recycled metals. Aluminum accounted for over 40% of the EE of the end product, whereas the use of recycled aluminum resulted in a significant reduction of this indicator. In addition, the use of recycled back cover reduced the GWPof the panel from 1.2×10^2 to 3.3×10^1 kg CO_2equiv. The findings also revealed that floating glass had a major contribution to the EE of the panel, corresponding to over 50% of the total EE. Overall, Scenario 4 resulted in the best possible and least energy-intensive solution.

Martinopoulos et al. (2013) assessed the environmental impacts from the use of different materials or manufacturing techniques in the production of DSHWS in Greece by implementing LCA. The environmental performance of the DSHWS analyzed was presented in two stages; firstly, by comparing the environmental impact of the systems during all life cycle stages, excluding utilization, and secondly, including utilization. The authors concluded that the systems presenting the lowest environmental impact during the first stage were not the same as those that presented the highest environmental gain during their whole life cycle. This was due to a much higher impact of substituted electricity in the use phase, which exceeded the small differences in the other stages. Moreover, as thermal efficiency differed from system to system, their environmental performance was influenced mainly by the conventional energy substituted and, to a lesser extent, by the materials used for their production.

Michael and Selvarasan (2017) compared a solar photovoltaic thermal (PV/T) based water heating system with a solar photovoltaic (PV) system and a flat plate DSHWS based on economic evaluation and environmental assessment in India. The results indicated that the solar thermal system had the best economic, energy, and environmental performance over the solar PV system and solar PV/T system. In detail, the collector cost occupied 41.8%, 37.2%, and 35.1% of the total cost of the solar thermal system, solar PV, and solar PV/T system, respectively. In contrast, the solar thermal system exhibited the lowest EPBT and simple payback period. Additionally, the corresponding values for the EE of the collector were 46%, 81.5%, and 72.6% of the total EE. Regarding the total CO_2 emissions, the collector's part of the total was 40.7%, 61.8%, and 60.8% in the solar thermal, solar PV, and solar PV/T system, respectively. Furthermore, the extra cost of including batteries in a stand-alone PV system increased the payback period in excess of 10 years and the benefit to cost ratio to less than 1.

Uctug and Azapagic (2018) implemented an LCA for a flat plate thermosiphon DSHWS in four regions with different climatic conditions in Turkey. For end-of-life treatment, two different scenarios were considered: in the first one, all the components were assumed to be landfilled, whereas, in the second, the recyclable components (steel, aluminum, and glass) were recycled. The obtained results were compared to the impacts from the use of a natural gas-based boiler. The results indicated that the construction stage was the most energy-intensive and had the most significant environmental impacts, while among the components, the water storage tank had the highest overall impact. In addition, the energy output was significantly higher than the input (7–28 times) and the payback periods ranged between 1 and 3 years. The findings also suggested that the environmental impacts of the systems were 1.5–2 times lower than those of the natural gas boilers in the hotter regions. Moreover, recycling the major components instead of landfilling reduced human toxicity potential by 50% but had only a small effect of 5% on the other impacts. Hence, they appeared high sensitivity to the type of material used for the storage tank assembly but not an important contribution of transport and end-of-life recycling. The only exception to the latter is human toxicity potential, which decreased significantly with more remarkable recycling.

Saint et al. (2019) reported the environmental impact of a unique integrated collector-storage solar water heater (ICSSWH) design, experimentally tested under Scottish weather conditions, considering the circular economy aspect and the reuse potential life cycle perspective. The analysis was completed for four scenarios, that is, an ICSSWH incorporating either fins or a baffle plate, each one evaluated with and without a circular approach. The results showed that the EE was not significantly affected for the circular against the typical linear approach. The corresponding improvements were 2.8% and 2.6% for the finned and baffled systems, respectively. When compared against the operational energy savings, payback times of 4.5–4.6 years for a circular and a linear approach, respectively, were recorded. As for the EC, the improvements of circular against the linear model were 47% and 45% for the finned and baffled collectors, respectively. Using the operational carbon savings, which far outstripped the EC, carbon payback times of just 7 months could be realized when replacing an electric system.

3.4 Evacuated tube solar thermal collectors

Milousi et al. (2019) evaluated the environmental profile of flat plate and vacuum tube collectors through a detailed cradle-to-grave LCA. Both systems were installed on existing buildings for residential applications in Greece. The outcomes of this analysis revealed that the production phase of the collector component is the leading process in terms of environmental burdens accounting for 57% and 45.3% (of all total inflows and outflows) for the flat plate and vacuum tube collector, respectively. In order to stress the environmental benefits and drawbacks of each collector type, a comparison between the two systems was conducted. Both collectors exhibited quite close environmental impacts in most categories, with the vacuum tube system having the highest values in most cases. An exception was the cumulative CO_2equiv., where the values ranged between 2.22×10^{-2} and 2.38×10^{-2} kg CO_2equiv./$kWh \cdot m^2$, and the lowest value corresponded to the vacuum tube collector.

Greening and Azapagic (2014) investigated the life cycle environmental impacts of flat plate and evacuated tube systems for DWH in the United Kingdom. They compared them to water heating provided by alternative means: a natural gas condensing boiler, electricity, and heat pumps. The results indicated that the flat plate collector is preferable as it exhibited on average 7% lower impacts than the evacuated tube system. Both solar thermal collectors had lower impacts than the gas boiler for 5 out of the 11 categories, performed better than the electrical water heating for 8 of the 11 categories, and were also more sustainable in comparison to the different types of heat pumps (air-ground and water-source), for 7 out of 11 impacts considered. Nevertheless, solar thermal systems do not necessarily represent an environmentally more sustainable alternative to fossil fuel-based water heating. This happens due to the lack of suitable locations in the United Kingdom and their limited energy output, which leads to the requirement of a backup heating system, typically a gas boiler.

Hang et al. (2012) studied the energetic, environmental, and economic impacts of a flat plate and an evacuated tube solar collector with auxiliary heaters (electricity, natural gas) in three different locations in the United States; Los Angeles, Atlanta, and Chicago. The performance of the solar systems was compared to the conventional water heating system in each city. The results showed that flat plate DSHWS with natural gas auxiliary heater had the best performance among all types and locations. On the other hand, the payback periods varied from 1 to 2 months, 1 to 4 months, and 4 to 13 years for CED, carbon footprint, and life cycle cost, respectively, for different cities and different configurations when using the conventional electrical water heating system in each city as the benchmark.

3.5 Conclusions

This chapter provides a preview of how LCA has been applied to solar thermal systems in the past decade and what lessons can be learned from the large group of LCA studies. The review provided herein is not intended to be extensive because of the large amount and diversity of researched systems. Nevertheless, it brings sufficient insights to realize that the application of some basic methodological steps could be better developed. For example, a comprehensive analysis of the system life cycle, for example, including the often-ignored disposal or decommissioning stage and all environmental impacts related to the studied systems, for example, not just focusing on climate change or energy-related issues should be better addressed in future research.

Nomenclature

BIST	building-integrated solar thermal
BoP	balance of plant
CED	cumulative energy demand
CO_2	carbon dioxide
DHW	domestic hot water
DSHWS	domestic solar hot water systems
EC	embodied carbon
EE	embodied energy
EF	ecological footprint
EPBT	energy payback time
GHG	greenhouse gas
GPBT	greenhouse gas payback time
GWP	global warming potential
ICSSWH	integrated collector-storage solar water heater
IS	indicator of sustainability
LCA	life cycle assessment
LCIA	life cycle impact assessment
PCM	phase change material
PED	primary energy demand
PV	photovoltaic
PV/T	photovoltaic Thermal
R2R	roll to roll
T	transport

References

Alberti, J., Raigosa, J., Raugei, M., Assiego, R., Ribas-Tur, J., Garrido-Soriano, N., et al., 2019. Life cycle assessment of a solar thermal system in Spain, eco-design alternatives and derived climate change scenarios at Spanish and Chinese National levels. Sustainable Cities and Society 47, 101467.

Altun-Çiftçioğlu, G.A., Gökulu, O., Kadırgan, F., Kadırgan, M.A.N., 2016. Life cycle assessment (LCA) of a solar selective surface produced by continuous process and solar flat collectors. Solar Energy 135, 284–290.

Arnaoutakis, N., Milousi, M., Papaefthimiou, S., Fokaides, P.A., Caouris, Y.G., Souliotis, M., 2019. Life cycle assessment as a methodological tool for the optimum design of integrated collector storage solar water heaters. Energy 182, 1084–1099.
Comodi, G., Bevilacqua, M., Caresana, F., Pelagalli, L., Venella, P., Paciarotti, C., 2014. LCA analysis of renewable domestic hot water systems with unglazed and glazed solar thermal panels. Energy Procedia 61, 234–237.
Greening, B., Azapagic, A., 2014. Domestic solar thermal water heating: a sustainable option for the UK? Renewable Energy 63, 23–36.
Hang, Y., Qu, M., Zhao, F., 2012. Economic and environmental life cycle analysis of solar hot water systems in the United States. Energy and Buildings 45, 181–188.
Kalogirou, S., 2009. Thermal performance, economic and environmental life cycle analysis of thermosiphon solar water heaters. Solar energy 83 (1), 39–48.
Koroneos, C.J., Nanaki, E.A., 2012. Life cycle environmental impact assessment of a solar water heater. Journal of Cleaner Production 37, 154–161.
Lamnatou, C., Notton, G., Chemisana, D., Cristofari, C., 2014. Life cycle analysis of a building-integrated solar thermal collector, based on embodied energy and embodied carbon methodologies. Energy and Buildings 84, 378–387.
Lamnatou, C., Notton, G., Chemisana, D., Cristofari, C., 2015. The environmental performance of a building-integrated solar thermal collector, based on multiple approaches and life-cycle impact assessment methodologies. Building and Environment 87, 45–58.
Lamnatou, C., Cristofari, C., Chemisana, D., Canaletti, J.L., 2016. Building-integrated solar thermal systems based on vacuum-tube technology: Critical factors focusing on life-cycle environmental profile. Renewable and Sustainable Energy Reviews 65, 1199–1215.
Lamnatou, C., Motte, F., Notton, G., Chemisana, D., Cristofari, C., 2018a. Building-integrated solar thermal system with/without phase change material: Life cycle assessment based on ReCiPe, USEtox and Ecological footprint. Journal of Cleaner Production 193, 672–683.
Lamnatou, C., Motte, F., Notton, G., Chemisana, D., Cristofari, C., 2018b. Cumulative energy demand and global warming potential of a building-integrated solar thermal system with/without phase change material. Journal of Environmental Management 212, 301–310.
Lamnatou, C., Cristofari, C., Chemisana, D., Canaletti, J.L., 2019. Payback times and multiple midpoint/endpoint impact categories about Building-Integrated Solar Thermal (BIST) collectors. Science of The Total Environment 658, 1039–1055.
Martinopoulos, G., Tsilingiridis, G., Kyriakis, N., 2013. Identification of the environmental impact from the use of different materials in domestic solar hot water systems. Applied Energy 102, 545–555.
Michael, J.J., Selvarasan, I., 2017. Economic analysis and environmental impact of flat plate roof mounted solar energy systems. Solar Energy 142, 159–170.
Milousi, M., Souliotis, M., Arampatzis, G., Papaefthimiou, S., 2019. Evaluating the environmental performance of solar energy systems through a combined life cycle assessment and cost analysis. Sustainability 11 (9), 2539.
Moore, A.D., Urmee, T., Bahri, P.A., Rezvani, S., Baverstock, G.F., 2017. Life cycle assessment of domestic hot water systems in Australia. Renewable Energy 103, 187–196.
Morsink-Georgali, F.Z., Kylili, A., Fokaides, P.A., 2017. Life cycle assessment of flat plate solar thermal collectors. Journal of Sustainable Architecture and Civil Engineering 21 (4), 41–49.

Piroozfar, P., Pomponi, F., Farr, E.R., 2016. Life cycle assessment of domestic hot water systems: a comparative analysis. International Journal of Construction Management 16 (2), 109–125.
Saint, R.M., Pomponi, F., Garnier, C., Currie, J.I., 2019. Whole-life design and resource reuse of a solar water heater in the UK. Proceedings of the Institution of Civil Engineers-Engineering Sustainability 172 (3), 153–164. Thomas Telford Ltd.
Uctug, F.G., Azapagic, A., 2018. Life cycle environmental impacts of domestic solar water heaters in Turkey: The effect of different climatic regions. Science of the Total Environment 622, 1202–1216.

CHAPTER 4

Environmental assessment of wind turbines and wind energy

Angeliki Kylili
Department of Environment, Ministry of Agriculture, Rural Development and Environment, Cyprus

Contents

4.1 Introduction

Wind energy is a key player in the global renewable energy generation sector and has a leading role in the transition towards a climate-neutral and sustainable world. Energy generated by renewable energy resources, such as wind, offers added value to the environment, the society, and the economy by reducing greenhouse gases emissions, lessening the environmental impact of the energy industry, improving air quality, and creating green job positions. Even though wind energy technologies do not emit emissions directly during their operation and the electricity generated is considered "green," there are however environmental burdens related to the sourcing of materials, the manufacture of the technologies' components, the assembly and installation of the infrastructure, and the decommissioning at their end-of-life. Accordingly, one has to look beyond their operational phase, in which the most benefits arise, and take into consideration their entire life

Environmental Assessment of Renewable Energy Conversion Technologies
DOI: https://doi.org/10.1016/B978-0-12-817111-0.00004-8

cycle. Understandably, the requirements for materials and energy for the construction and assembly of these enormous structures are substantial. Especially, when considering the case of offshore and remote wind parks, the transportation of materials to the site, in terms of fuel and energy consumption, also adds to the environmental burden of the overall project. Keeping in mind the research and innovation trends and the fact that the wind energy sector is taking steps toward bigger-scale, offshore, and floating systems, the accurate calculation of wind turbines' environmental performance and the establishment of environmentally optimal designs become critical, if the goal is the creation of climate-neutral societies, based on circular economy models.

Within this context, holistic methodologies and tools that can account for both the environmental benefits and burdens throughout the life cycle of product systems, have a major role in fulfilling the energy, climate, and sustainability targets. Life cycle assessment (LCA) is one of these methodologies. It currently represents the most established tool for the scientific community, the industry, and the policy-makers for assessing the environmental impact of any product or system. This chapter focuses on the environmental assessment of wind turbines and wind energy, based on LCA methodologies. Firstly, the latest research trends of the wind energy sector are introduced before a comprehensive overview of the life cycle inventory (LCI) and LCA studies for wind turbines and wind energy is conducted. These parts of the chapter provide key parameters that are considered in the implementation of LCAs of such systems, as well as present significant findings and figures on their environmental impact. Concluding, a critical review is undertaken on the LCA studies reviewed in this work and significant challenges in the environmental assessment of wind turbines and wind energy are identified.

4.2 State-of-the-art on wind turbines and wind energy

Since the latter half of the 20th century, the field of exploitation of wind for electricity generation has experienced substantial growth, with great leaps taken in the development of wind turbine designs. Wind turbines are realized based on the orientation of the shaft and rotational axis, with both horizontal-axis wind turbines (HAWT) and vertical-axis wind turbines (VAWT) having their own favorable characteristics and limitations (Table 4.1). Although the HAWT is the dominant design configuration, extensive research also surrounds the VAWT for future development.

Table 4.1 Advantages and limitations of wind turbines technologies.

Wind turbines category	Wind turbines technologies	Advantages	Limitations
Horizontal-axis (HAWT)	Upwind Downwind	▪ Turbines exposed to stronger winds due to height of the tower ▪ Higher efficiency due to the blades moving perpendicular to wind	▪ Heavy tower construction to support blades, gearbox, and generator ▪ Lift of components to the top of the tower for positioning during installation ▪ Down-wind designs can cause metal fatigue and structural failure (up-wind designs are preferred) ▪ Yaw control mechanism is required to turn the blades toward the wind and stop spinning in high winds to prevent damage ▪ Cyclic stresses and vibrations can cause damage to the blade roots, hub, and axle of the turbines ▪ Can be installed only in rural areas

(Continued)

Table 4.1 (Continued)

Wind turbines category	Wind turbines technologies	Advantages	Limitations
Vertical-axis (VAWT)	Darrieus rotor Giromill Savonius rotor Twisted Savonius	■ Generator and transmission devices located at ground level—easier access for maintenance purposes ■ Capture wind from any direction ■ Lower production costs ■ Yaw drive and pitch mechanism not required ■ Easy installation and transportation ■ Can be installed in urban- and low-lying areas or other areas where wind is less predictable ■ Low risk for human and birds due to lower blade speeds	■ Lower efficiency due to reduced energy capture and additional drag ■ Small motor required to provide initial push to start ■ Guy-ropes that provide support to the rotor take up large land extensions ■ Noise pollution ■ Increased maintenance costs due to bearing wear caused by vibration

In fact, research and innovation in the wind energy sector are continuous, pushing the wind technologies higher in the technology readiness levels. The end goal of the wind energy sector is to develop breakthrough technologies that will facilitate the integration of the generated renewable electricity in the grid, increase the reliability and quality of the energy generated, and decrease the costs of wind energy. The realization of this goal means that wind energy becomes competitive with conventional fossil-fueled energy. By addressing specific technical and horizontal issues and focusing research toward a better understanding of the interrelations

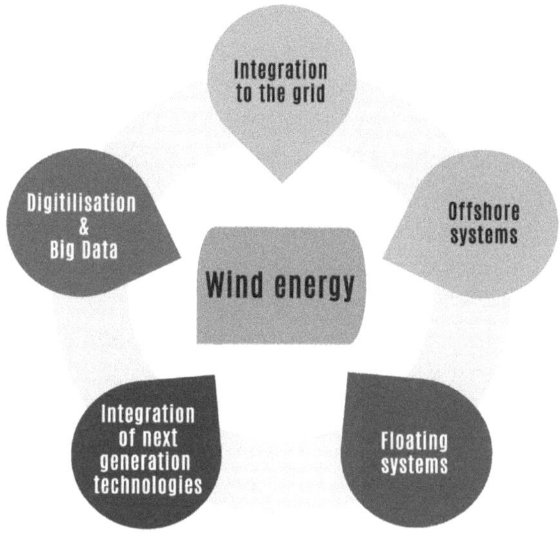

Figure 4.1 State-of-the-art research in the wind energy sector.

between them, innovative solutions and cutting-edge technologies for the wind sector can be delivered. This section elaborates on the current state-of-the-art surrounding the wind sector and presents the research priorities of the industry (Fig. 4.1).

One of the key issues of the wind industry—common among all renewables technologies' sectors—is the integration of the energy generated by the plants into the power systems. Strategic and technological advancements in the energy system will enable the development of more resilient grids that are able to handle the variable nature of renewable electricity in higher amounts. In particular, research focuses on developing innovative solutions that will provide the energy system with more flexibility, such as the development of advanced virtual power plants and substation concepts. Additionally, research efforts are made for the improvement of storage solutions that offset the variability of wind energy and maintain balance in the overall system. Novel hybrid systems, including wind-solar, wind-pumped hydro, and wind-thermal, are also envisioned to play a significant role in the near-future decentralized energy systems of variable demand and supply. The development of hybrid technologies will enable the provision of necessary features and services that enhance the reliability and stability of the system. Furthermore, strategic advancements in grid planning and operation will also need to accompany the technological solutions, especially in the cases of offshore wind plants, which will allow a wider penetration and exploitation of the generated renewable electricity.

Another key aim of the wind sector is to move from onshore to offshore wind energy generation, which directly relates to achieving further cost reductions in the offshore designs and technologies. Achieving these cost reductions entails not only new developments for increasing the size and power output of wind turbines, but is also critical for research in the planning for the connection of the offshore wind power plants to the electricity grid and the layout and operation of wind power plants. Accordingly, the state-of-the-art research focuses on analyzing the challenges relevant to offshore grid expansion and DC connection costs, such as more synergies between wind power plants and grid can be achieved. The significance of providing feasible solutions on this aspect is highlighted by the fact that the offshore balance of plants accounts for more than 50% of the development costs (Hundleby and Freeman, 2017). Directly related to this, is also the development of the offshore electrical infrastructure, including the analysis of intercable layout configurations, installation of DC connections, and the optimal location of substations (ETIPWind, 2018). As already mentioned, cost reductions will be achieved by increasing the scale of economies; that is building larger offshore wind turbines that will produce greater power outputs, more energy- and cost-efficient. This not only comprises an economic assessment of bottom fixed and floating systems, but also the design of the appropriate substructures to support the larger turbines. The energy efficiency of the wind turbines will also be facilitated through the development and employment of multiscale environmental modeling tools that integrate the wind, wave and soil interaction and better predict the wind energy potential.

The wind industry has particularly shifted its focus towards floating wind technologies due to the large opportunities offered in terms of industrialization and scalability. The research underpinning the floating wind market regards the development of holistic system designs, the creation of synergies between different sectors, and the establishment of appropriate regulatory frameworks and supply chains that will unlock the technology's potential and facilitate its market uptake and widespread deployment. Within this context, the interactions between wind turbines, foundations, and station-keeping systems are analyzed, such as lighter, more robust, and more resource-efficient designs can be delivered. Additionally, the coupling of the floating wind sector with ocean fish farms for water treatment and the oil and gas sector is anticipated to increase the economic viability of offshore wind. Furthermore, a prerequisite for the widespread application of the technology is the preparation of the market. For this purpose, the first step is the identification of the challenges related to the development of the technology for establishing a supporting regulatory framework, as well as an appropriate-to-technology stable supply chain of installation and hoisting systems. The need for the development of deployment models, case studies, and market assessments for large-scale deployment, installation, and operation of floating wind farms in various markets and environments is also identified.

The targets set for the offshore wind power generation also imply the need for increased employment of advanced technologies and tools, such as automation, machine learning, and next-generation robotics, to enable the smart operation of wind power plants. The installation and use of smart devices and intelligent sensors can enhance the operation and maintenance (O&M) of the wind farms by increasing its performance through better energy flows management with the use of automatic big data collection and analysis and by reducing the maintenance costs upon the capability of inspecting the condition of critical system components remotely. The decrease in the costs of wind energy O&M will also be brought upon the development of comprehensive approaches for the plants' lifetime extension. A better understanding of the individual components and overall performance through engineering solutions, digital solutions, and human resources will enable lifetime management and lifetime extension. Additionally, the coupling of sensor data from external conditions with the control system facilitates the analysis of the performance and health of wind turbines and their components to increase yields and reduce failure rates.

Digitalization and big data management will contribute to the transformation of the wind energy sector beyond the state-of-the-art, offering extensive opportunities for improvements. The data collected from wind turbines can be analyzed, such as the turbines' designs are optimized and the wind power plants' O&M strategies are improved through machine learning technologies. The availability of data also enables the development and validation of sophisticated models and big data tools that will have the capacity to analyze a large number of data and variables for improved accuracy and reliability of wind turbine performance. Additionally, research in the field focuses on the use of next-generation materials, such as nanomaterials, self-healing materials, novel coating materials that protect against edge erosion, and superior materials that reduce fatigue failure of the system components by making them lighter, more robust, lower-cost, and more sustainable structures. Within the same context, the need for developing industry-wide strategies to reuse and recycle raw materials and rare-earth minerals is also highlighted.

Accordingly, significant technological leaps in material recycling processes are required, which embrace the concept of a circular economy for the maximization of the energy- and resource-saving potential and the minimization of the sector's environmental impact.

4.3 Life cycle inventory of wind turbines and wind energy

The rate of development of renewable energy technologies (RETs), including wind, is rapidly affecting the validity and reliability of LCIs used in LCA studies directly. Accordingly, LCIs of wind turbines need to be regularly updated to incorporate the most recent and up-to-date data, such as consistent and representative results can be obtained (Turconi et al., 2013). Considering a cradle-to-gate approach, the development of LCIs of wind turbines and wind energy incorporates the data compilation for upstream, operational, and downstream processes, as well as the inclusion of the foreground and background systems. The relevant upstream processes include raw material extraction, manufacturing, and wind turbine construction, the operational includes power generation, operation, and maintenance, and the downstream includes decommissioning and end-of-life scenarios. The foreground system concerns the wind plants' construction and operation, while the background system refers to all the production processes for the entering and the exiting flows (Mendecka and Lombardi, 2019). The typical life cycle stages taken into consideration within the scope in LCA studies of wind turbines are illustrated in Fig. 4.2.

Similar to all LCA studies, the LCIs for the development of LCA studies of wind turbines and wind energy use both primary and secondary data. Primary data regarding energy and material inflows and outflows of the product system is collected for the upstream, operation, and downstream processes and the foreground system. This data can be obtained from the relevant industry professionals in the form of raw data or from reliable literature sources, whereas according to the International Organization for Standardization (ISO) 14040 series, any assumptions made should also be clearly stated. For instance, the development of LCI data regarding the transportation of the different components of wind turbines commonly involves the adoption of assumptions during the calculation of transportation distances using maps (Cámara et al., 2010; Mendecka and Lombardi, 2019). Where primary data is not available, secondary data, typically regarding inventory data for electricity supply or the sourcing and the production of materials, are retrieved from well-established LCI databases. The work of Davidsson et al. (2012) specifies the two other main

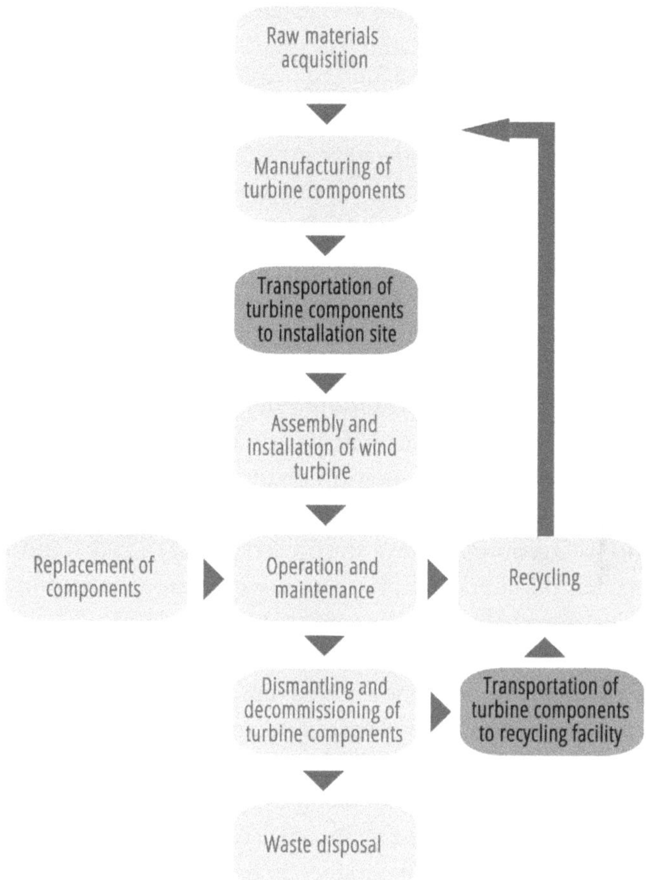

Figure 4.2 Life cycle stages considered in the implementation of Life Cycle Assessment studies of wind turbines and wind energy.

approaches that can be used for the development of LCI of wind energy systems. The input–output (I/O) analysis employs economic data for the estimation of the resource use in different parts of the economy, where an I/O matrix is developed, incorporating in a quantified way the inter-industrial relationships such as the interactions between different economic sectors are represented. The second approach is the process chain analysis (PCA), which estimates the actual physical flows of mass and energy. This is the approach based on which the largest percentage of studies are conducted. It is worth mentioning that studies using attributional approaches based on PCA indicate considerably lower energy use for operations and maintenance than studies that use I/O analysis (Davidsson et al., 2012).

In relation to the use of LCI databases for the implementation of LCAs of wind turbines and wind energy, ecoinvent represents the most established database and is widely used in the majority of the reviewed studies (Table 4.2). The article of Treyer and Bauer (2016a,b) elaborates on the electricity production datasets included in the ecoinvent database. With reference to the electricity production with wind turbines, ecoinvent categorizes the production into four sets in relation to capacity and location: for onshore electricity production, the categories are capacities of <1 MW, 1–3 MW, and >3 MW, while for offshore electricity production, one category of 1–3 MW capacities. The shares of the four categories were determined based on data for individual turbines installed in August 2011 in countries with wind power generation (TheWindPower, 2011). Furthermore, it is stated that the location-specific yield or capacity factor is critical in the implementation of LCA studies of wind energy.

A major point of uncertainty in the LCA of wind turbines and wind energy is the background process the LCI data (Mendecka and Lombardi, 2019). The scientific literature proposes novel approaches and methodologies to address this challenge. Mendecka and Lombardi (2019), following a comprehensive review and analysis of LCA results conducted for wind turbines of variant capacities, developed simplified LCA models based on harmonization procedures and regression analysis, while also identifying their advantages and limitations against the detailed approach. The novelty of the proposed approach is that the number of input parameters to the model is significantly decreased compared to the conventional approach, while a satisfactory level of uncertainty is maintained. Under the context of these simplified LCA models, the main input LCI data for production processes are treated as variables, such as deterministic input secondary data are converted into probabilistic data. By integrating temporal and spatial information into the LCI process, the work of Maier et al. (2017) established a dynamical LCI framework that generates detailed spatiotemporal results and therefore improves the reliability of the results. More specifically, spatialized LCI overtime was attained by incorporating into the LCI framework the enhanced structural path analysis (ESPA) method combined with spatial analysis. The effectiveness and reliability of the proposed methodology was demonstrated through its application to a dataset of electricity generation from 5 MW offshore wind turbines. Within the context of this methodology, a simplified LCI is developed by using the relevant ecoinvent dataset and making assumptions and simplifications on the process flows. As a result, the dynamical LCI framework enables the

Table 4.2 Overview of LCA studies for wind turbines and wind energy.

Study	Scope	Object(s) of investigation	Functional unit	System boundaries	Lifetime	Employed LCA software/databases	LCIA methodology	Sensitivity analysis
Ardente et al. (2008)	Define energy and environmental performance of wind electricity production	Onshore farm comprised of 11 wind turbines of 660 kW nominal power of each	1 kWh of electricity	Cradle-to-grave	20 years	ANPA; Boustead Consulting	n/r	Yes
Martínez et al. (2009a)	Investigate environmental impact of a wind turbine with DFIG	2 MW rated power onshore wind turbine	1 kWh of electricity	Cradle-to-grave	20 years	SimaPro	Eco-indicator 99	No
Martínez et al. (2009b)	Define possible environmental improvements for a wind turbine with DFIG	2 MW rated power onshore wind turbine	1 kWh of electricity	Cradle-to-grave	20 years	SimaPro	CML 2000	Yes
Tremeac and Meunier (2009)	Perform LCAs of a high power and a small wind turbines	4.5 MW wind turbine (for wind farm) and 250 W wind turbine (building or isolated site installations)	1 kWh of electricity	Cradle-to-grave	20 years	SimaPro	n/r	Yes
Weinzettel et al. (2009)	Compare LCA results of electricity generation between offshore	Floating farm consisting of 40 wind	1 MJ of electricity	cradle-to-grave	21 years	SimaPro; Ecoinvent	CML 2 baseline 2000	No

(Continued)

Table 4.2 (Continued)

Study	Scope	Object(s) of investigation	Functional unit	System boundaries	Lifetime	Employed LCA software/databases	LCIA methodology	Sensitivity analysis
	floating wind power plants and conventional offshore non-floating wind power plants	turbines of 5 MW nominal power each						
Martínez et al. (2010)	Assess the degree of uncertainty of LCA study of a multi-megawatt wind turbine due to choices made	2 MW rated power onshore wind turbine	1 kWh of electricity	Cradle-to-grave	20 years	SimaPro	Eco-indicator 99	Yes
Lahuerta and Saenz (2011)	Evaluation of different waste scenarios for composite parts (rotor and nacelle) of wind turbine	All wind turbines installed in mainland Spain between 1978 to early 2008	1 GWh of energy	Cradle-to-grave	20 years	SimaPro; Ecoivent	CML 2 baseline	Yes
Zhong et al. (2011)	Compare the environmental impacts of a polycrystalline PV module and a wind turbine	600 kW onshore wind turbine	1 kWh of electricity	Cradle-to-grave	20 years	SimaPro; Ecoinvent	Eco-indicator 99	No
Demir and Taşkın (2013)	Assess the environmental impact of 3 medium- and 2 large- scale wind turbines with variable hub heights	Onshore wind turbines of 330 kW, 500 kW, 810 kW, 2050 kW, and 3020 kW rated power	1 kWh of electricity	Cradle-to-grave	20 years	GaBi	CML	No

Garrett and Rønde (2013)	Assess the environmental impact of 2 MW grid streamer turbines	50 MW onshore wind plant comprised of 2 MW wind turbines	1 kWh of electricity	Cradle-to-grave	20 years	GaBi; Worldsteel; EUROFER; PlasticsEurope; EAA	CML 3.6	Yes
Rashedi et al. (2013)	Compare LCA results of 3 wind farms: onshore horizontal, offshore horizontal, and vertical axis.	50 MW wind firms comprised of 5 MW turbines of rated power for horizontal-axis and 1.6 MW for vertical axis	1 kWh of electricity	Cradle-to-grave	20 years	SimaPro; Ecoinvent	Eco-indicator 99	No
Haapala and Prem preeda (2014)	Compare the environmental impact of two 2 MW wind turbine designs	2 MW onshore wind turbine	2 MW wind turbine	Cradle-to-grave	20 years	SimaPro	ReCiPe 2008	Yes
Siddiqui and Dincer (2017)	Compare the environmental impacts of nuclear, wind, and hydro power generation in Canada	Onshore wind farm cpmrised of 2 MW turbines	1 kWh of electricity	Cradle-to-grave	20 years	GaBi	CML 2001	Yes
Ozoemena et al. (2018)	Assess the potential for advances in technology to increase the performance of a 1.5-MW wind turbine	114 MW onshore wind farm comprised of 1.5 MW rated power turbine	1 kWh of electricity	Cradle-to-grave	25 years	Ecoinvent	CML 2001	No

(Continued)

Table 4.2 (Continued)

Study	Scope	Object(s) of investigation	Functional unit	System boundaries	Lifetime	Employed LCA software/databases	LCIA methodology	Sensitivity analysis
Xu et al. (2018)	Study the environmental performance of utility-scale wind power systems in China	49.5 MW onshore wind farm comprised of 18 1.5 MW and 30 750 kW turbines	1 kWh of electricity	Cradle-to-grave	20 years	GaBi; Ecoinvent	CML 2002	Yes
Alsaleh and Sattler (2019)	Consuct LCA for large onshore wind turbines in the US	400 MW onshore wind farm comprised of 200 2 MW turbines	1 kWh of electricity	Cradle-to-grave	20 years	SimaPro	TRACI	Yes
Sacchi et al. (2019)	Propose a novel approach to build a comprehensive parameterized model that generates unique wind turbine LCIs conditioned by technologically, temporally and geographically-sensitive parameters	Sets of 100 kW, 500 kW, 1 MW, and 2 MW onshore wind turbines and 2 MW offshore wind turbines	1 kWh of electricity	Cradle-to-grave	20 years	LCA_WIND_DK; Ecoinvent	Proposed metho dogy	Yes
Wang et al. (2019)	Compare the environmental impacts of three power generation methods: hydro-, nuclear and wind power	49.5 MW wind farm comprised of 1.5 MW rated power turbines	1 kWh of electricity	Cradle-to-grave	20 years	SimaPro	CML 2001	Yes

implementation of a more accurate Life Cycle Impact Assessment (LCIA). This can indicate environmental impact overtime, as well as provide results on a local level, given that the appropriate local-specific inventory data is introduced into the methodology. A similar approach for the development of a parameterized inventory for wind turbines—by taking into consideration changes over space and time of specific background processes in the LCI—was also adopted by Sacchi et al. (2019). This study performed an analysis of 1401 cradle-to-grave LCI of wind turbines. It examined into detail the variability in results for the environmental performance of wind turbines based on significant influencing factors including nominal power output, onshore or offshore installation, diversity of designs and material intensity, wind availability, service time, and time of turbine manufacture. Presented with these novel approaches in the development of inventories for wind turbines and wind energy, the potential of adopting more sophisticated parameterized LCIs that incorporate technological, spatial, and temporal variability is demonstrated. This will not only improve the reliability of the inventories but also allow the implementation of LCA studies that provide more accurate results regarding the environmental performance of wind energy projects.

4.4 Life cycle assessment of wind turbines and wind energy

4.4.1 Key parameters in the implementation of life cycle assessment studies

A comprehensive review of the existing scientific literature on the LCA for wind turbines and wind energy indicated the identification of typical and commonly-found parameters that are used when implementing LCA, including the functional unit (FU), the system boundaries, and the lifetime of wind turbines, as well as the use of energy- and carbon dioxide- related indicators and payback indices (Table 4.2). In particular, this analysis revealed that the chosen FU in this field of research is one unit of generated electricity, namely kilowatt hour (kWh), Gigawatt hours (GWh), and megawatt (MW), with 1 kWh of electricity being the most common across studies. Regarding system boundaries, all studies conduct cradle-to-grave LCAs due to the significance of the final disposal and waste management stages in wind turbines' overall energy and environmental performance. Typically, a lifetime duration of 20 years for the wind turbines is assumed based on the manufacturers' turbine efficiency guarantee notes. However, a lifetime of 25 years has also been considered by a recent study (Ozoemena et al., 2018).

It is evident that the use of such standard parameters in the implementation of LCA studies also facilitates the comparative assessment between the variant designs and capacities of wind turbines that currently exist, as they conclude to comparable energy and environmental impact indicators. Although studies report for energy-related indicators, such as energy intensity and cumulative energy demand and carbon-related indicators including carbon dioxide-equivalent (CO_2-equiv.), intensity, and embodied carbon, the most indicative and commonly used index in the LCA of wind turbines is the energy payback period. Table 4.3 presents energy- and carbon-related indicators obtained from the implementation of past relevant LCA studies. The energy intensity ranges between 0.015 kWh_{prim}/kWh_e, reported for a floating offshore wind farm, to 0.33 kWh_{prim}/kWh_e, reported for a 250 W turbine. The respective range for carbon dioxide–equivalent (CO_2-equiv.) intensity is 7.2 g CO_2-equiv./kWh_e, provided for a 2.0 MW wind turbine and 56 g CO_2-equiv./kWh_e for a 100 kW wind turbine (Table 4.3). There is a common agreement among the scientific community that the energy payback time (EPBT) of the majority of wind turbines and wind energy is less than a year, proving to be an energy-efficient electricity generation technology considering its 20-year guaranteed lifetime. It is noteworthy that sensitivity analysis is not performed or reported in all the implemented studies, although the implementation of such an analysis is strongly recommended in the relevant ISO standards. The work of Martínez et al. (2010) provided significant findings on the degree of uncertainty of an LCA study conducted for a multi-megawatt wind turbine due to different choices that have been made in relation to maintenance, needs of material and energy used, the percentage of recycled materials during the disposal, waste treatment of the wind turbine, and the composite waste treatment of the blades. The sensitivity analysis indicates 14%–20% variations in the LCA results, mainly arising from variations in the maintenance and recycling of wind turbine's components.

4.4.2 Significant findings from previous life cycle assessment studies

The object of investigation of the majority of studies is the definition of the energy and environmental performance of novel turbine designs for detecting improvement in opportunities or the comparison between different designs. Table 4.4 presents an overview of the LCIA results extracted from past scientific studies, illustrating the potential of the wind turbines or wind parks under investigation in specific environmental

Table 4.3 Energy- and carbon-related indicators resulted from the implementation of LCA studies of wind turbines and wind energy.

Study (year)	Location	Power capacity (MW)	Energy intensity (kWh_{prim}/kWh_e)	Carbon dioxide–equivalent intensity (g $CO_{2\text{-equiv.}}$/kWhe)	Payback index (years)
Ardente et al. (2008)	Italy	7.26	0.040–0.070	8.8–18.5	<1.0
Martínez et al. (2009a)	Spain	2.0	—	—	0.4
Martínez et al. (2009b)	Spain	2.0	0.019	—	0.6
Tremeac and Meunier (2009)	France	4.5 0.00025	0.08 0.33	15.8 46.4	0.5 1.5
Weinzettel et al. (2009)	Norway	200.0	0.015	—	0.4
Demir and Taşkın (2013)	Turkey	0.33 0.5 0.81 2.05 3.02	— — — — —	15.1–38.3	2.5–3.0[a] 2.3–2.9[a] 1.52[a] 1.3–1.5[a] 1.7–2.3[a]
Garrett and Rønde (2013)	—	50.0	—	7.2–9.3	0.7–0.9
Haapala and Prempreeda (2014)	USA	2.0	—	—	0.4 (M1); 0.5 (M2)
Ozoemena et al. (2018)	UK	114.0	—	10.3–16.6	—
Xu et al. (2018)	China	49.5	—	—	0.5
Alsaleh and Sattler (2019)	USA	400.0	—	—	2.5
Sacchi et al. (2019)	Denmark	0.1 0.5 1.0 2.0	— — — —	45.3 19.1 21.6 17.3	— — — —

M, Model.
[a]Represents the estimated values from published figures.

Table 4.4 Potential environmental impact as resulted from the implementation of LCA studies of wind turbines and wind energy.

Study	GWP	AP	EP	ADP (elements)	ADP (fossils)	FAETP	HTP	MAETP	ODP	POCP	TETP
	kg CO_2-equiv.	kg SO_2-equiv.	kg PO_4-equiv.	kg Sb-equiv.	MJ	kg DCB-equiv.	kg DCB-equiv.	kg DCB-equiv.	kg CFC11-equiv.	kg C_2H_4-equiv.	kg DCB-equiv.
Ardente et al. (2008)	1.48E-02								6.59E-10	1.61E-05	
Martínez et al. (2009a)	6.58E-03	5.43E-05	5.68E-06	3.75E-05		2.81E-03	1.55E-05	4.41E + 00	5.21E-10	2.13E-06	1.56E-04
Weinzettel et al. (2009)	8.89E-04	8.89E-06	9.44E-07	6.67E-06		8.89E-04	6.39E-03			4.72E-07	1.81E-05
Demir and Taşkın (2013)	1.63E-02 4.04E-02	5.78E-05 1.27E-04	5.39E-06 1.27E-05			3.63E-05 7.18E-05	1.05E-03 2.02E-03			5.97E-06 1.38E-05	2.12E-05 4.82E-05
Garrett and Rønde (2013)	7.70E-03	3.70E-05	3.70E-06	4.40E-07	1.00E-01	1.00E-04	1.15E-03	1.10E + 00		4.10E-06	1.90E-05
Siddiqui and Dincer (2017)	1.21E-02	6.85E-05	8.20E-06				7.90E-05			3.30E-06	

Ozoemena et al. (2018)	1.18E-02	9.17E − 05	6.90E − 05	8.91E − 05		1.95E − 02	5.38E − 02	4.48E + 01	1.24E − 09	6.54E − 06	2.24E − 04
Xu et al. (2018)	8.65E-03	7.25E-05	5.45E-05	2.49E-07	9.34E-02	3.33E-02	5.46E-02	4.78E + 00	3.68E-08	6.36E-06	1.30E-03
Alsaleh and Sattler (2019)	1.80E − 02	1.04E − 04			2.02E − 02				2.69E − 09		
Wang et al. (2019)	2.83E-02	2.02E-04	9.45E-05				3.15E-02			1.66E-05	

GWP, Global warming potential; *AP*, acidification potential; *EP*, eutrophication potential; *ADP(elements)*, abiotic depletion of elements potential; *ADP(fossils)*, abiotic depletion of fossil fuels potential; *FAETP* freshwater aquatic ecotoxicity potential; *HTP*, human toxicity potential; *MAETP*, marine aquatic ecotoxicity; *ODP*, ozone depletion potential; *POCP*, photochemical ozone creation potential; *TETP*, terrestrial ecotoxicity potential.

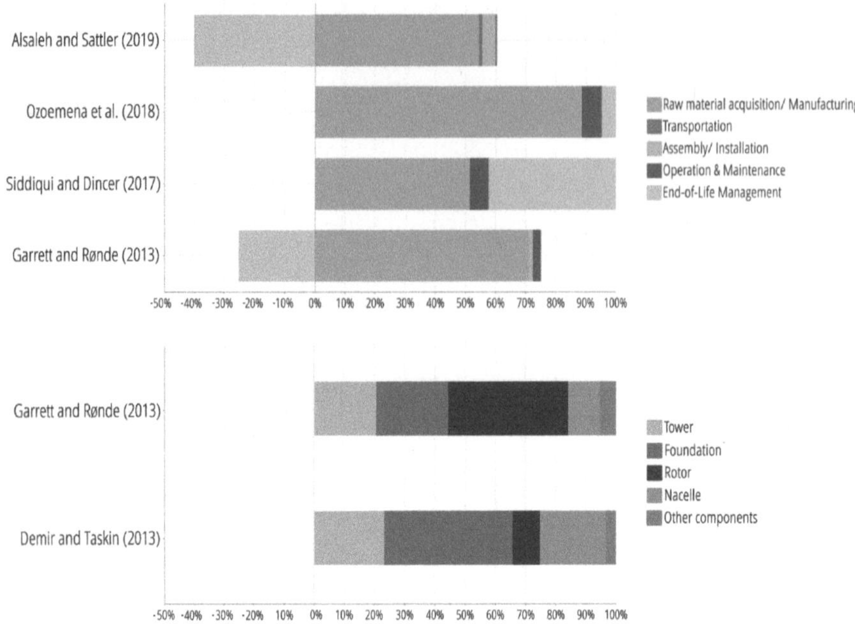

Figure 4.3 Contribution of life cycle stages and turbine components to the overall potential and global warming potential of wind turbines.

impact categories. Fig. 4.3 demonstrates the contribution of the different life cycle stages and the different system components to the overall global warming potential.

The focus of Martínez et al. (2009a,b) was an onshore wind turbine with a doubly-fed inductor generator (DFIG) with 2 MW rated power. The results of the studies indicated that the use of cement causes the most ecological effects of inorganic respiration. At the same time, the non-recycling of the turbine blades is the main contributor to the wind turbine's environmental impact. Although their impact is 89% to 99% smaller, depending on the environmental category, in comparison to conventional power plants, there is a need for improvement in the manufacturing processes, the recycling opportunities, and the practical application in the dismantling and waste management life cycle stages. Tremeac and Meunier (2009) used the LCA methodology to assess the environmental impact of two wind turbines of different rated power—a 4.5 MW and 250 W wind turbines, in support of wind energy generation. The findings highlighted the significance of the high efficiency of turbines with relation to the potential of wind on the chosen site of installation, the proximity of the components' manufacturing and installation sites, as well as the recyclability of the materials during decommissioning. Using detailed industry-published LCI datasets, Garrett and Rønde (2013) performed LCA for 2 MW turbines, which indicated an EPBT of 8–11 months and recyclability of 81%–85% of turbine mass (Table 4.1). This study also discusses further applications of LCA, with the purpose of improving the product internally for external communications. A total of five LCA studies have been conducted by Ozoemena et al. (2018) for a 1.5 MW wind turbine to identify the better performing design based on technology improvements. The results of the Global warming potential (GWP) analysis indicated that there is no consistent relationship that can be established between wind turbine capacity and contribution to environmental impacts even though, considering equivalent FUs, the GWP is typically lower in larger-scale turbines than smaller-scale turbines. Due to its significance in wind turbines' overall life cycle performance, the final disposal and waste management phase was the main focus of the study of Lahuerta and Saenz (2011). Four different waste scenarios for the rotor and nacelle were examined to determine their environmental impact and effect on the EPBT of the whole system. This study indicated that improvement in the potential environmental impact in the range of 5%–10% could be achieved from recycling the components rather than being sent to landfill or incinerated, with similar improvements (10%) also found in the EPBT results. The work of Demir and Taşkın (2013) examined the environmental effects of three medium-scale and two large-scale wind turbines of 330, 500, 810 kW and 2050, 3020 kW, respectively, with hub heights of 50, 80, and 100 m. The finding of the LCA studies showed that the EPBT of the turbines with high hub heights is quicker due to increased electricity generation (Table 4.1). The comparison of the environmental impact of two wind turbines of equivalent rated power (2.0 MW) was the key objective of the study of Haapala and Prempreeda (2014). The LCA results confirmed previous findings on the manufacturing stage's significant contribution, while the turbine components' transportation distances also influence the environmental impact. The one model was found to have shorter EPBT, mainly due to differences in the design of the tower, where one of the models required more steel than the other.

A few studies employ LCA methodologies for the comparative environmental assessment of different technologies of wind turbines and other renewable technologies, extending the investigation beyond the wind energy sector. In an investigation of the life cycle environmental impact of three wind farms, comprised of different wind generation technologies, one onshore horizontal, one offshore horizontal, and one vertical axis, it was indicated that the vertical axis wind farm has the lower impact per unit of generated electricity (Rashedi et al., 2013). The same study also concludes that wind farms are demonstrating the second-best environmental performance for electricity generation, following the hydro-kinetic technologies, among a selection of conventional and RETs, including viz, oil, diesel, coal, natural gas, solar, biomass, and nuclear. These findings are found in agreement with a previous work, where the environmental impact of a wind turbine and a polycrystalline photovoltaic (PV) module were compared, modeling landfill disposal and recycling LCA scenarios. The superior performance of wind turbines is attributed to the greater potential for recycling material as a result of a larger volume of waste generated at the decommissioning stage, compared to solar PV modules, where smaller quantities of waste become available.

Directly relevant to the scope of this book chapter is also the implementation of LCAs for wind energy generation, where the study broadens its system boundaries also to include the manufacturing and construction of additional system components, building structures, and auxiliary facilities required for the development of wind farms, such as transformers and transformer stations, lay-bays, foundations, cable trenches, path connections, etc. As it is evident from Table 4.1, the environmental impact of wind farms is found among the lower part of the range, with the work of Ardente et al. (2008) reporting for energy intensity of 0.04 kWh_{prim}/kWh_e and CO_2-equiv. the intensity of 8.8 CO_2-equiv./kWh_e and the work of Weinzettel et al. (2009) for 0.015 kWh_{prim}/kWh_e and EPBT of 5, 2 months, considering the case of a floating offshore wind farm.

4.5 Critical review on the environmental assessment of wind turbines and wind energy

A critical review of the existing literature of environmental assessment of wind energy technologies undertaken in this work has led to the identification of a number of challenges for the LCA practitioners and the scientific community of the relevant field. Some of the challenges identified are horizontal, shared challenges common across almost any type of product or product system assessed using LCA methodologies, whereas others are technology-specific (Fig. 4.4).

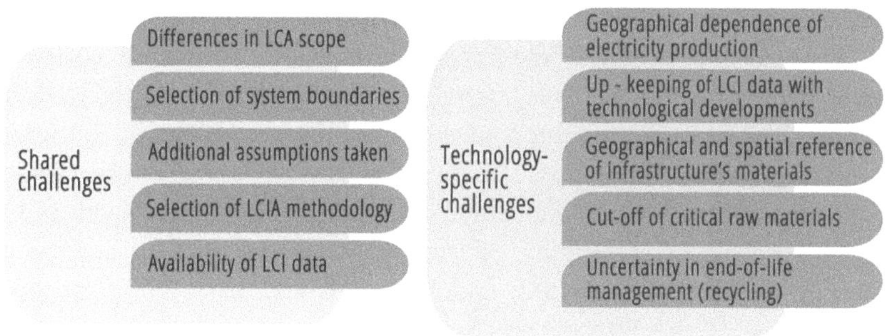

Figure 4.4 Challenges in the life cycle assessment (LCA) of wind turbines and wind energy.

4.5.1 Shared challenges related to life cycle assessment

Common challenges across several fields of LCA application include the choice of assumptions, methodologies, and the phase of data collection for the development of LCIs. The implementation of LCI and LCA studies follows a standardized approach, based on the international standards ISO 14040 and ISO 14044, which distinguish the studies into separate phases with specific guidelines on what needs to be included and what needs to be taken into consideration in each phase. Starting from the first phase, the Goal and Scope Definition phase, a difficulty identified is the direct comparison of environmental impacts among a pool of wind energy technologies. Although the majority of studies use energy- and carbon dioxide-related indicators and payback indices to enable as far as possible a comparison between different studies, yet the scope of each study is different, as the results are intended for use for a specific type of audience, or they serve or support a specific goal. Taking this into account, it is challenging, for instance, to assess the breadth of environmental impact reductions as a result of technological improvements or technical improvements in the specifications of wind turbines solely by relying on the findings of an LCA study (Elsam, 2004; Haapala and Prempreeda, 2014). Directly related to the scope of the LCA is the selection of system boundaries to be taken into consideration that define the development of a "cradle-to-cradle," "cradle-to-grave," "cradle-to-gate," or "gate-to-gate" type of study. Additional assumptions taken into consideration and choices made by LCA practitioners such as the selection of FU, cut-off limits, expected lifetime, and LCIA methodology can also have major impacts on the final results (Ardente et al., 2008). However, this review has indicated that, for the case of LCI and LCA wind turbines' studies, more typically, the same FU (1 unit of generated electricity) and same boundary conditions (cradle-to-grave) are employed, as well as equal expected wind turbines' lifetimes (20 years) are assumed (Table 4.2). Nevertheless, in comparing baseload and intermittent power technologies, such as wind turbines, the comparison on equal terms is problematic given the different levels of reliability in the electricity supply of the two technologies. To address this challenge, Gagnon et al. (2002) and Fthenakis and Kim (2007) proposed the modeling of the intermittent power technology as one of a storage system, such as the services provided by the two systems are technically the same (Turconi et al., 2013). Regarding the choice of LCIA methodology for the calculation of the results on the potential environmental impact, it represents a typical limitation that makes difficult the intercomparison of studies' results or their use for decision-making, since indisputable values for the environmental effects produced by a considered system cannot be generated (Mendecka and Lombardi, 2019). Some researchers even debate whether the results of LCA studies that employ different LCIA methodologies should be included under the same comparison framework at all (Davidsson et al., 2012; Haapala and Prempreeda, 2014). However, it is noteworthy that some of the limitations related to the choices made during the first phase of LCA can be directly addressed through the implementation of a sensitivity analysis, as indicated for a number of studies in Table 4.2, where the extent of the impact of a particular choice can be assessed and defined in quantified terms. Relating to the second phase of an LCA study is the data collection for the development of the LCI. Despite this being another typical challenge for LCA practitioners of any field, several studies highlight the specific difficulties in collecting data for the implementation of LCA for wind turbines. This is justified as wind turbines are made up of many components and sub-components of different functionalities and therefore of different mechanical, electrical, and electronic parts, and acquiring or collecting all this type of information and data from the suppliers is not only difficult, but also very time-consuming (Martínez et al., 2009b; Cámara et al., 2010; Ozoemena et al., 2018; Rashedi et al., 2013).

4.5.2 Technology-specific challenges related to life cycle assessment

Again, in the case of technology-specific challenges identified through the implementation of this comprehensive analysis, the majority of them can be found within the first phase of the LCA methodology. Considerable challenges arise from the technological, geographical, and temporal scope of LCA studies of wind turbines. The most contributing factor for these challenges is the nature of the technology itself, which is very site-depended, as is the case for all renewable energy systems (Ardente et al., 2008). In terms of wind speed and frequency, the wind potential depends on the geographic location and area characteristics, and it is directly related to the operation efficiency of the wind energy system. In turn, these factors, which cannot be considered constant (average values can only be used), have a great impact on the subsequent calculation of LCA indices such as payback time and energy- and carbon-related intensities. Additionally, the reference year and the geographical origin of the materials and energy used for the infrastructure have also been identified as factors that can greatly influence the LCA results (Turconi et al., 2013). The weighting of this, is of particular significance, especially in the case of wind energy technologies and other rapidly growing alternative energy technologies, which are subject to constant technological advancements. This signifies the importance of the upkeeping of the relevant LCIs with the fast pace of the sector's technological development and their simultaneous development and updating with the most recent data to enhance LCA results' reliability (Edenhofer et al., 2011; Turconi et al., 2013). This type of limitation has been specifically pointed out in the publications of Treyer and Bauer (2016a,b) on the LCIs of electricity generation and power supply in the well-established LCI database, ecoinvent. These publications suggest the LCIs, which do not accurately represent the latest developments in specific regions and specific technologies, which are not available in some countries, can result in overestimated values of environmental impact from the electricity generation of those markets. The selection of cut-off criteria has also been identified as of particular importance in LCA studies of clean energy technologies due to the use of rare earth elements and special metals for the construction of their infrastructure. It is a fact that rare earth elements, such as neodymium in wind turbines, represent critical raw materials for the technologies, although they are not included in the majority of LCAs as they exist in very small quantities compared to the total mass of the structures. Another reason for this type of element being excluded from LCAs, as provided by Davidsson et al. (2012), is that the use of nonenergy resources is not of relevance in LCA studies. Yet, regardless of the reason, challenges arise from this, especially when considered in a wider environmental and circular economy perspective, since rare earth elements are almost always associated with radioactive heavy metals and their availability in the market is considered to become "high risk" in the near future (Kanazawa and Kamitani, 2006; Davidsson et al., 2012). As discussed earlier in this section, the calculations of energy- and carbon-related indicators for wind energy technologies incorporate some level of inaccuracies since these systems are site-specific, and the amount of electricity generated is largely dependent on the local wind potential. This fact gives rise to another challenge with regard to the variability between studies of the methods adopted for the estimation of the annual production of electricity of a wind turbine or a wind farm for the calculation of intensity indicators, which are represented by the ratio between either the primary energy consumed or the CO_2 emissions and the electricity production during the wind turbine lifetime (Tremeac and Meunier, 2009). In order to calculate electricity production, a capacity factor is used that is the ratio between the actual electricity generated and the potential electricity to be generated when the system operates continuously at its rated power (Davidsson et al., 2012). However, it has been identified that some of the reviewed studies provide a capacity factor (Weinzettel et al., 2009; Demir and Taşkın, 2013; Ozoemena et al., 2018; Rashedi et al., 2013; Haapala and Prempreeda, 2014) or use measured data of turbine performance and wind plant losses (Garrett and Rønde, 2013), where in other studies the approach used for the calculation of the indices is not clearly explained (Ardente et al., 2008; Martínez et al., 2009a,b; Tremeac and Meunier, 2009; Lahuerta and Saenz, 2011). For that reason, Davidsson et al. (2012) characterize the use of such indicators as "problematic" and propose the use of high and low production estimates rather than the use of fixed production rates. The same study supports the use of energy return on energy investment and EPBT for the definition of the energy performance of wind turbines in LCA studies, provided that the calculations of these indicators are based on standardized methodologies. A notable point of reference in the LCA studies of wind turbines is the great environmental benefit of the recycling phase on the overall life cycle environmental impact of the systems; the recycling of the tower and nacelle has been indicated to reduce the total impact by 52% and 31%, respectively (Martínez et al., 2009a,b; Martínez et al., 2010). The recycling of the materials that make up the structure of wind turbines, metals specifically (steel, copper, aluminum, cast iron), whose at least 90% of the mass is considered to be recycled at the final stage, reduces the environmental impact significantly and produces negative values for this phase. In the study of Zhong et al. (2011), the total impact in the category of carcinogens is even presented to be negative due to recycling. This implies that the construction of a wind turbine decreases carcinogenic effects, which in reality can be questioned. Another question that arises in relevance to the recycling phase is whether the materials of wind turbines will be actually recycled in a 20-year period, considering the projected high costs of decommissioning, especially in the cases of offshore or remote wind parks (Davidsson et al., 2012; Rashedi et al., 2013). It is established that the recycling of wind turbines during decommissioning is crucial for good environmental impact figures, while the disposal of wind turbines without recycling represents a worst-case scenario in terms of environmental impacts (Tremeac and Meunier, 2009; Zhong et al., 2011).

Concluding, it is also important to mention that LCA methodologies have also been criticized for not addressing major environmental issues; additional environmental burdens that are relevant to wind power include land use and noise. However, in terms of land usage requirements and the amount of energy to be extracted in relation to the land area, wind turbines have been proven to perform well in comparison to conventional power plants, overruling the claims for excessive land use by wind energy (Hau, 2006).

References

Alsaleh, A., Sattler, M., 2019. Comprehensive life cycle assessment of large wind turbines in the US. Clean Technologies and Environmental Policy 21 (4), 887–903.

Ardente, F., Beccali, M., Cellura, M., Brano, V.L., 2008. Energy performances and life cycle assessment of an Italian wind farm. Renewable and Sustainable Energy Reviews 12 (1), 200–217.

Cámara, E.M., Macías, E.J., Fernández, J.B., de la Parte, M.P., 2010. Environmental impact of modern wind power under LCA methodology. INTECH Open Access Publisher, London.

Davidsson, S., Höök, M., Wall, G., 2012. A review of life cycle assessments on wind energy systems. The International Journal of Life Cycle Assessment 17 (6), 729–742.

Demir, N., Taşkın, A., 2013. Life cycle assessment of wind turbines in Pınarbaşı-Kayseri. Journal of Cleaner Production 54, 253–263.

Edenhofer, O., Pichs-Madruga, R., Sokona, Y., Seyboth, K., Matschoss, P., Kadner, S., et al., 2011. IPCC special report on renewable energy sources and climate change mitigation. Prepared By Working Group III of the Intergovernmental Panel on Climate Change, Cambridge University Press, Cambridge, United Kingdom.

Elsam, 2004. Life Cycle Assessment of Offshore and Onshore Sited Wind Farms. Elsam Engineering A/S, Denmark.

ETIPWind (European Technology & Innovation Platform on Wind Energy), 2018. Strategic research and innovation agenda 2018. <https://etipwind.eu/> (accessed 11.09.18).

Hundleby, G., Freeman, K., 2017. Unleashing Europe's Offshore Wind Potential: A New Resource Assessment. BVG Associates, Swindon, United Kingdom.

Fthenakis, V.M., Kim, H.C., 2007. Greenhouse-gas emissions from solar electric-and nuclear power: a life-cycle study. Energy Policy 35 (4), 2549–2557.

Gagnon, L., Belanger, C., Uchiyama, Y., 2002. Life-cycle assessment of electricity generation options: the status of research in year 2001. Energy Policy 30 (14), 1267–1278.

Garrett, P., Rønde, K., 2013. Life cycle assessment of wind power: comprehensive results from a state-of-the-art approach. The International Journal of Life Cycle Assessment 18 (1), 37–48.

Haapala, K.R., Prempreeda, P., 2014. Comparative life cycle assessment of 2.0 MW wind turbines. International Journal of Sustainable Manufacturing 3 (2), 170–185.

Hau, E., 2006. Chapter 15: Environmental impact. Wind Turbines: Fundamentals, Technologies, Application, Economics. Springer Science & Business Media, Germany.

Kanazawa, Y., Kamitani, M., 2006. Rare earth minerals and resources in the world. Journal of Alloys and Compounds 408, 1339–1343.

Lahuerta, F., Saenz, E., 2011. Life cycle assessment of the wind turbines installed in Spain until 2008. In: Proceedings of the Europe's Premier Wind Energy Conference and Exhibition. Brussels, Belgium, pp. 14–17.

Maier, M., Mueller, M., Yan, X., 2017. Introduction of a spatiotemporal Life Cycle Inventory method using a wind energy example. Energy Procedia 142, 3035–3040.

Martínez, E., Jiménez, E., Blanco, J., Sanz, F., 2010. LCA sensitivity analysis of a multi-megawatt wind turbine. Applied Energy 87 (7), 2293–2303.

Martínez, E., Sanz, F., Pellegrini, S., Jiménez, E., Blanco, J., 2009a. Life cycle assessment of a multi-megawatt wind turbine. Renewable Energy 34 (3), 667–673.

Martínez, E., Sanz, F., Pellegrini, S., Jiménez, E., Blanco, J., 2009b. Life-cycle assessment of a 2-MW rated power wind turbine: CML method. The. International Journal of Life Cycle Assessment 14 (1), 52–63.

Mendecka, B., Lombardi, L., 2019. Life cycle environmental impacts of wind energy technologies: a review of simplified models and harmonization of the results. Renewable and Sustainable Energy Reviews 111, 462–480.

Ozoemena, M., Cheung, W.M., Hasan, R., 2018. Comparative LCA of technology improvement opportunities for a 1.5-MW wind turbine in the context of an onshore wind farm. Clean Technologies and Environmental Policy 20 (1), 173–190.

Rashedi, A., Sridhar, I., Tseng, K.J., 2013. Life cycle assessment of 50 MW wind firms and strategies for impact reduction. Renewable and Sustainable Energy Reviews 21, 89–101.

Sacchi, R., Besseau, R., Pérez-López, P., Blanc, I., 2019. Exploring technologically, temporally and geographically-sensitive life cycle inventories for wind turbines: a parameterized model for Denmark. Renewable Energy 132, 1238–1250.

Siddiqui, O., Dincer, I., 2017. Comparative assessment of the environmental impacts of nuclear, wind and hydro-electric power plants in Ontario: A life cycle assessment. Journal of Cleaner Production 164, 848–860.

TheWindPower, 2011. Wind turbines and windfarms database. <www.thewindpower.net> (accessed 11.09.18).

Tremeac, B., Meunier, F., 2009. Life cycle analysis of 4.5 MW and 250 W wind turbines. Renewable and Sustainable Energy Reviews 13 (8), 2104–2110.

Treyer, K., Bauer, C., 2016a. Life cycle inventories of electricity generation and power supply in version 3 of the ecoinvent database—Part I: Electricity generation. The International Journal of Life Cycle Assessment 21 (9), 1236–1254.

Treyer, K., Bauer, C., 2016b. Life cycle inventories of electricity generation and power supply in version 3 of the ecoinvent database—Part II: Electricity markets. The International Journal of Life Cycle Assessment 21 (9), 1255–1268.

Turconi, R., Boldrin, A., Astrup, T., 2013. Life cycle assessment (LCA) of electricity generation technologies: overview, comparability and limitations. Renewable and Sustainable Energy Reviews 28, 555–565.

Wang, L., Wang, Y., Du, H., Zuo, J., Li, R.Y.M., Zhou, Z., et al., 2019. A comparative life-cycle assessment of hydro-, nuclear and wind power: A China study. Applied Energy 249, 37–45.
Weinzettel, J., Reenaas, M., Solli, C., Hertwich, E.G., 2009. Life cycle assessment of a floating offshore wind turbine. Renewable Energy 34 (3), 742–747.
Xu, L., Pang, M., Zhang, L., Poganietz, W.R., Marathe, S.D., 2018. Life cycle assessment of onshore wind power systems in China. Resources, Conservation and Recycling 132, 361–368.
Zhong, Z.W., Song, B., Loh, P.E., 2011. LCAs of a polycrystalline photovoltaic module and a wind turbine. Renewable Energy 36 (8), 2227–2237.

CHAPTER 5

Environmental assessment of biomass thermochemical conversion routes through a life cycle perspective

Kyriakos Panopoulos[1], Giorgos Kardaras[1,2], Tzouliana Kraia[1] and Michael Bampaou[1]
[1]Chemical Process and Energy Resources Institute (CPERI), Centre for Research and Technology Hellas (CERTH), Thessaloniki, Greece
[2]Department of Mechanical Engineering, University of Western Macedonia, Greece

Contents

5.1 Introduction

According to the Renewable Energy Directive 2018/2001/EC, the greenhouse gas (GHG) emission savings from the use of biofuels, bioliquids, and biomass fuels shall be (Article 29, paragraph 10).

Environmental Assessment of Renewable Energy Conversion Technologies
DOI: https://doi.org/10.1016/B978-0-12-817111-0.00007-3

- At least 50% for biofuels, biogas consumed in the transport sector, and bioliquids produced in installations in operation on or before October 5, 2015.
- At least 60% for biofuels, biogas consumed in the transport sector, and bioliquids produced in installations starting operation from October 6, 2015 until December 31, 2020.
- At least 65% for biofuels, biogas consumed in the transport sector, and bioliquids produced in installations starting operation from January 1, 2021.
- at least 70% for electricity, heating, and cooling production from biomass fuels used in installations starting operation from January 1, 2021 until December 31, 2025 and 80% for installations starting operation from January 1, 2026.

Thermochemical conversion is one of the routes to convert biomass into renewable end products. Pyrolysis and gasification are among those thermochemical systems that aim to act as mitigation agents of GHG and reduce fossil fuel dependency.

Unfortunately, several issues concerning the sustainability and environmental benefit of thermochemical conversion pathways hinder their wide-scale implementation at a commercial level. These issues include the security of feedstock supply, energy efficiency, and GHG emissions savings, the technological barriers concerning biomass processing, and the environmental, economic, and social impacts associated with the utilization of natural resources. These matters can be summarized in the triptych "Environment-Economy-Society," often presented as three interrelated circles, with sustainability being placed at the intersection, as shown in Fig. 5.1.

Numerous sustainability assessment tools have been developed, such as life cycle assessment (LCA), substance flow analysis, sustainability impact assessment, environmental risk analysis, multicriteria analysis, environmental extended input—output analysis, and cost—benefit analysis.

LCA is the most widely used assessment tool to analyze the major environmental impacts of various products, to determine how changes in processes could lower the environmental impact, to compare the environmental impacts of different products, and ultimately, to evaluate the overall sustainability based on the environmental and socioeconomic impacts of a product, process, or pathway across its entire life cycle, including resource extraction, manufacture, use, reuse, recycle, and final disposal (Welcomer, 2008).

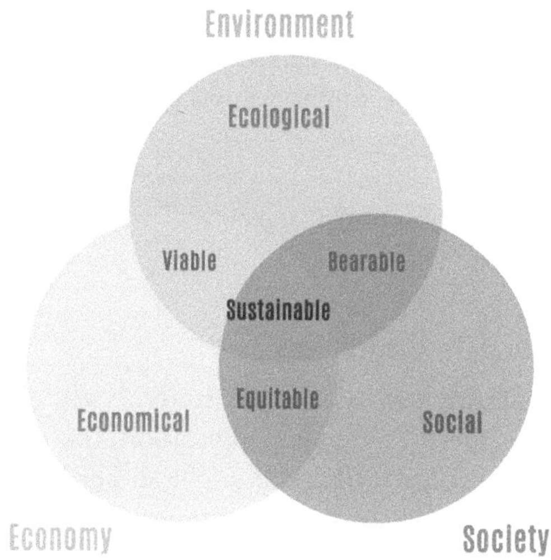

Figure 5.1 Interrelated circles of sustainability. *Adapted from Purvis, B., Mao, Y., Robinson, D., 2019. Three pillars of sustainability: in search of conceptual origins. Sustainability Science 14(3), 681–695.*

According to Cherubini and Strømman (2011), the comparison of LCA studies is complicated due to the variations in critical methodological choices and assumptions like goal and scope definition, input data, allocation methods, impact assessment methods, etc. These choices strongly affect the outcome of LCA studies, and even for similar conversion routes, results may differ.

5.2 Life cycle assessment of biomass conversion routes

The concept of life cycle thinking exceeds the approach of assessing a product system from only the manufacturing process and takes one step further to include the entire life cycle. According to ISO 14040, "life cycle" is defined as the "consecutive and interlinked stages of a product system, from raw material acquisition or generation from natural resources to final disposal," thus, making the LCA one of the most prominent tools for measuring the environmental and health impacts of goods and services by quantifying all relevant emissions and resources consumed along with their supply chain activities (from the extraction of raw materials through production, use, and recycle up to the final waste management).

In compliance with ISO 14040 (2006) and 14044 (2006), four phases shall be completed in carrying out an LCA study. These phases are distinct but interconnected within the methodological framework

- Goal and scope definition
- Life cycle inventory (LCI)
- Life cycle impact assessment (LCIA)
- Interpretation

5.2.1 Goal and scope definition

Goal definition is the first step of an LCA study. It sets the overall context for the study, specifically the intended application, the reasons for carrying out the study and decision context, the target audience, the commissioner of the study, and the identity of the practitioner.

The scope of an LCA study describes in detail the system to be evaluated and the technical specifications. The scope definition shall be in line with the defined goals of the study and include the functional unit (FU), the system boundary, assumption/limitations, impact assessment method, and relevant additional information such as technical aspects and/or physical properties of the product under investigation.

5.2.1.1 Functional unit

The FU is identified as a key aspect of LCA since it qualitatively and quantitatively describes the main function and duration of the product system in scope and ensures the comparability and interpretation of the results. The standards clearly state that the choice and formulation of FU must be consistent with the goal and scope of the study. According to Ahlgren et al. (2015), the FU is closely related to the study's aim and thus depends on the research question.

5.2.1.2 Boundaries

ISO 14040 (2006) and 14044 (2006) state that the boundary definition is an iterative process, meaning that initial system boundaries must be set based on the goal and scope of the study and the final system boundaries are determined after the initial calculations and the sensitivity analysis. The definition of system boundaries indicates the activities or processes included and those excluded from the analysis. Different system boundaries are used to assess the environmental impact of biomass conversion routes.

- "Cradle-to-gate or well-to-tank (for liquid biofuels)" is used only for comparing various conversion technologies. The use phase and disposal phase of the product are omitted from the analysis.
- "Cradle-to-grave or well-to-wheel" is an open-loop assessment of a product system's full life cycle that includes the material and energy production chain and all processes from the raw material extraction through the production, transportation, and use phase up to the product's end-of-life treatment.
- "Gate-to-grave" analysis includes the processes solely from the use and end-of-life phases of a product.
- "Gate-to-gate" analysis is used for assessing the environmental impacts of a single production step or process.
- "Cradle-to-cradle" includes a closed-loop analysis of the entire life cycle of a product system, where the end-of-life disposal phase for a product is a recycling process.

According to Muench and Guenther (2013), a one-dimensional definition of the system boundary (Fig. 5.2) has been quite heterogeneous and sometimes even contradictory. In biomass conversion systems, three paths play an important role in the definition of the system boundary (Fig. 5.3). The first path is biomass production, transportation, conversion, and end-of-life treatment. The second path is the construction and decommissioning of the conversion system. The third path is the distribution of energy. With this approach, the definitions discrepancies can be eliminated and enhance the comparability between LCA studies.

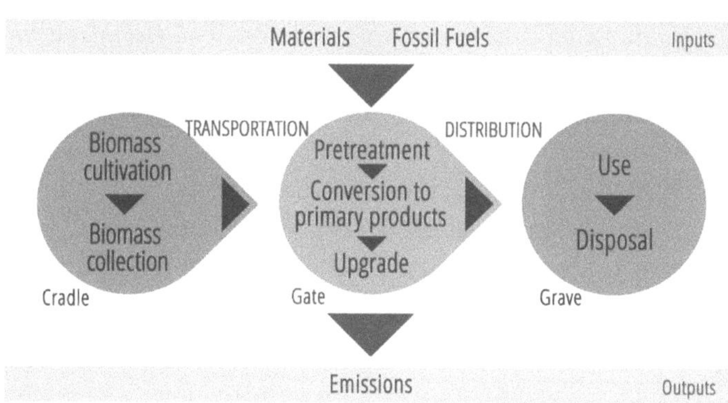

Figure 5.2 Schematic representation of a one-dimensional system boundary. *Adapted from (Sarangi et al., 2018).*

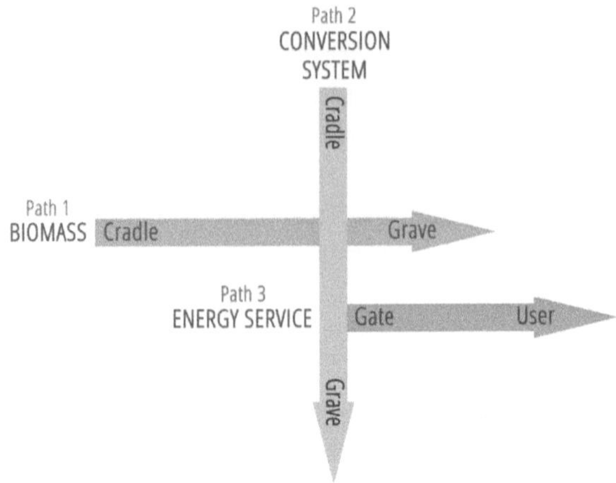

Figure 5.3 Biomass conversion systems pathways (Muench et al., 2013).

5.2.2 Life cycle inventory

The second stage in performing an LCA and, at the same time, the most demanding and time-consuming task is the collection of data related to the product system under investigation. In general, a product system consists of two types of processes:

1. *Foreground processes*: refer to processes that describe the production system under investigation.
2. *Background processes*: are processes for the production of generic materials, energy, transport, and waste management.

The distinction between these process types is not always clear and depends on the aim of the LCA study, that is, in the case of performing an LCA of thermochemical conversion of biomass, the truck transporting the biofuel to the end user can be considered as a background process. However, in the case of conducting an LCA about trucks using biofuel, the inputs and outputs that are specific to the trucks must be considered as foreground processes.

The ISO 14044 standard states that primary data shall be obtained for all foreground and background processes, where appropriate. However, if secondary data is more representative or appropriate than primary data (to be justified and reported) for foreground processes, secondary data shall also be used for the foreground processes. Secondary data should be used only for background processes.

Primary data: Data collected (measured, calculated, or estimated) from production sites associated with the processes within the system boundary.

Secondary data: Data derived from other sources such as literature or databases. For secondary data, the following data quality requirements should be specified (ISO 2006b):

- Time-related coverage
- Geographical coverage
- Technology coverage
- Precision
- Completeness
- Consistency
- Sources of the data
- Uncertainty of the information

5.2.3 Life cycle impact assessment

LCIA plays an integral and critical role in an LCA study. It is the phase during which the results of the LCI are assigned to relevant categories, and the potential environmental impact of the investigated system is assessed. According to ISO 14040/44 standard, LCIA shall consist of three obligatory stages and three optional, namely:

Obligatory stages:

- Selection of relevant impact categories. Impact categories must reflect environmental issues raised by the goal and scope of the study.
- Classification. Classification aims to assign the LCI results into impact categories according to their ability to contribute to environmental problems.
- Characterization. Each LCI result contributing to one impact category is multiplied with a characterization factor and then added to formulate the category indicator result.

Optional stages:

- Normalization. In normalization, the category indicator result is divided by a reference value to solve the incompatibility of units and enhance comparability.
- Grouping. Grouping allows sorting the impact categories into larger concern groups such as energy, transport, etc.
- Weighting. In the weighting step, the category indicator results are weighted and ranked based on factors.

Every LCA must at least include the obligatory stages. If these steps are not included, the study can be considered as an LCI. Furthermore, for comparative, external studies, weighting is not allowed according to the

ISO series. A vast number of standard impact assessment methods are available, developed by various institutions and researchers, that can be applied in an LCA study. There is no consensus on the best method, but efforts have been made for the evaluation and provision of guidance on this issue (ILCD handbook on LCIA). There are two main approaches for the classification and characterization of LCI results: the use of impact category indicators at the "endpoint" level (damage-oriented) and the use of indicators at the "midpoint" level (problem-oriented) that is between the LCI result and the "endpoint." In general, midpoint indicators express impacts into environmental matters such as global warming, ozone depletion, etc., and have lower uncertainty. On the other hand, endpoint indicators express areas of concern, like damage to human health, ecosystems, and resources. They are easier to understand and interpret by relevant stakeholders and decision-makers. Fig. 5.4 presents a typical life cycle impact pathway of thermochemical biomass conversion technologies.

5.2.4 Interpretation of results

The last of the four steps of an LCA is the interpretation of the impact assessment results. This is achieved by identifying the most relevant processes or significant issues based on the results of the LCI and LCIA. The main outcomes of this phase are conclusions and recommendations; limitations are also stated. This stage intends to help the audience understand the results and guide the relevant stakeholders in their decision-making process. The ISO 14044 standard describes a number of checks, such as completeness, sensitivity, and consistency, to test whether conclusions are adequately supported by the data and by the procedures used.

Completeness checks assess the LCI data to ensure that it is entirely relative to the defined goal, scope, and system boundary. Sensitivity checks assess the extent to which variations in assumptions, data, and methods affect the results. Consistency checks assess whether the data, assumptions, procedures, methodological choices, and models applied during the analysis are consistent with the goal and scope of the study. Any issues flagged in checks may result in the revision of the previous stages (LCI and LCIA) or the readjustment of the scope of the study.

The selection of the appropriate methodology and impact categories, as well as tackling key methodological issues (i.e., goal definition, system boundaries, FU, allocation of biomass feedstock and products, choice of

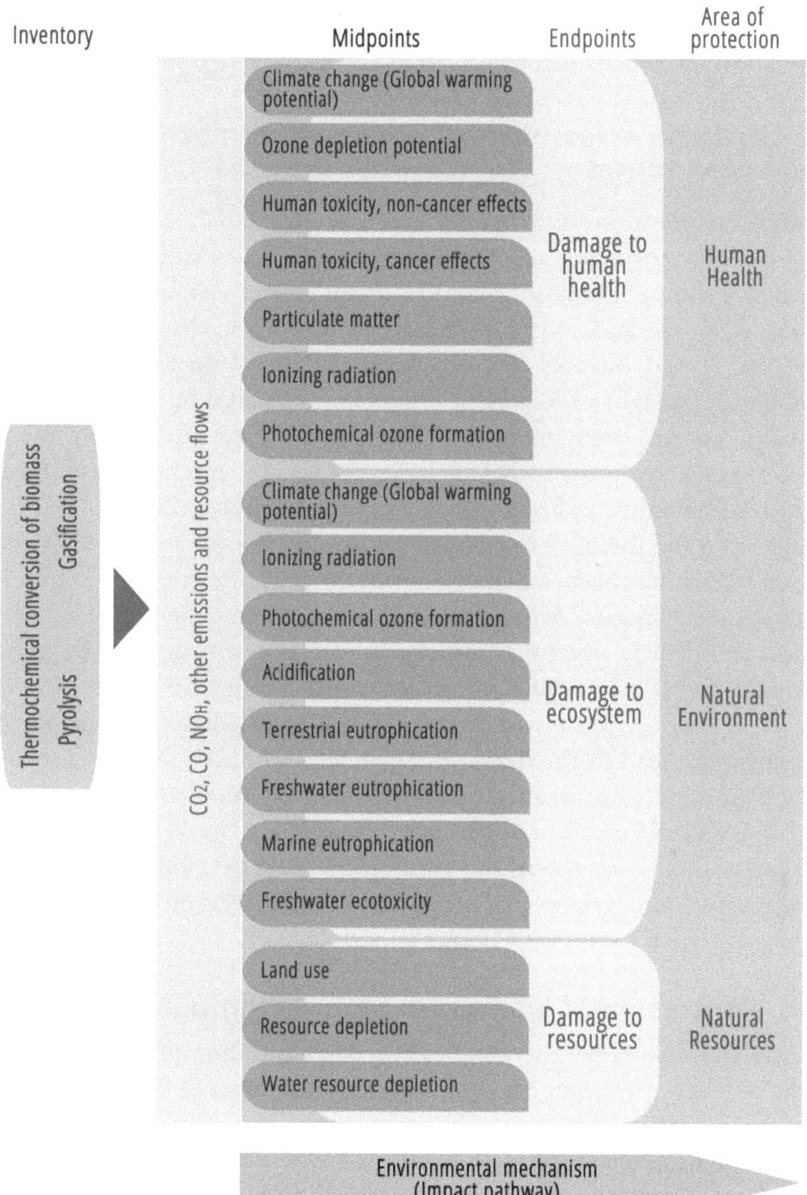

Figure 5.4 Example of life cycle impact pathway. *Adapted from the International Reference Life Cycle Data System (ILCD) Handbook, 2010.*

data, and land use, LU) in a harmonized way can enhance reproducibility and comparability of the LCA studies.

5.3 Life cycle assessment of biomass thermochemical conversion routes

Pyrolysis, gasification, and hydrothermal liquefaction are the three main processes of thermochemical conversion routes. They utilize heat to convert biomass into a variety of liquid and gaseous renewable end products, such as heat, electricity, transportation fuels, syngas, char, and chemicals. The quality and quantity of the products depend on the difference in operating temperatures, reaction, residence time, heating rate, amount of required oxygen, etc., used by each process. Their advantage over the biochemical conversion routes includes the shorter reaction time, greater feedstock flexibility, and higher operating temperatures (Brown, 2019).

Between the three, pyrolysis and gasification are the most widely studied, and LCA has been extensively used to quantify their environmental footprint. Applying LCA methodology to the thermochemical conversion routes can identify potential bottlenecks and hotspots during the process steps and reveal issues concerning the sustainability of the end products. The realization of the environmental benefits and drawbacks of biomass utilization can lead to the design of "greener" products. However, LCA is not bulletproof, and as already stated, different methodological choices tend to affect the results and the outcome of LCA studies. The next section focuses on the pyrolysis and gasification processes, followed by a literature review, uncovering the issues that affect the comparability of LCA studies and the robustness of LCA methodology.

5.3.1 Conversion of biomass to biofuels through pyrolysis

Biomass pyrolysis is the thermochemical decomposition process of biomass in the absence of oxygen. It is an endothermic process, requiring heat to split the biomass macromolecule constituents into useful fractions thermally. The three main pyrolysis products are bio-oil, biochar, and permanent gases. The yields and compositions of each product phase strongly depend on the employed operating parameters such as biomass composition (Oasmaa et al., 2010; Di Blasi et al., 1999; Tsai et al., 2006), feedstock particle size (Akhtar and Amin, 2012; Neves et al., 2011; Rezaei et al., 2016), residence times (Akhtar and Amin, 2012; Pütün et al., 2007a; Tripathi et al., 2016), and temperature and heating rates

(Beis et al., 2002; Uzun et al., 2006; Pütün et al., 2007b). For example, slow heating rates (0.1–2°C/s), long residence times (hours to days), and moderate temperatures (<500°C) favor the production of biochar (slow pyrolysis), whereas high heating rates (>10–200°C/s), short residence times (0.5–10 seconds), and temperatures around 500°C favor the production of bio-oil (fast pyrolysis) (Bridgwater, 2012; Fahmy et al., 2020). Since most of the reviewed LCA pyrolysis studies focus on the production of liquid fuels, the fast pyrolysis process is the preferred mode of operation. Fig. 5.5 illustrates a simplified block diagram of the biomass fast pyrolysis system.

Several pretreatment technologies should be employed, depending on the requirements and specifications of the employed pyrolysis reactor technology in the next stage. A critical factor is the contained moisture in the biomass feedstock (Jahirul et al., 2012). The moisture of the feedstock material, along with the produced water during the dehydration reactions of pyrolysis, would result in its accumulation in the liquid phase (Wang et al., 2017; Collard and Blin, 2014). This would, among others, result in a reduction of the energy content and the density of the bio-oil as well as affect catalysts if the bio-oil is further used in upgrading processes (Venderbosch et al., 2010). Therefore, drying of the biomass feedstock is required. In addition, particle size plays a vital role in effectively transferring heat to the biomass feedstock (Akhtar and Amin, 2012). Generally, small particles are preferred (<3 mm) for the faster and uniform heating of the feed (Bridgwater, 2012). Other feedstock parameters requiring additional pretreatment measures are the ash content, the biomass constituents, and the chemical nature of the feedstock.

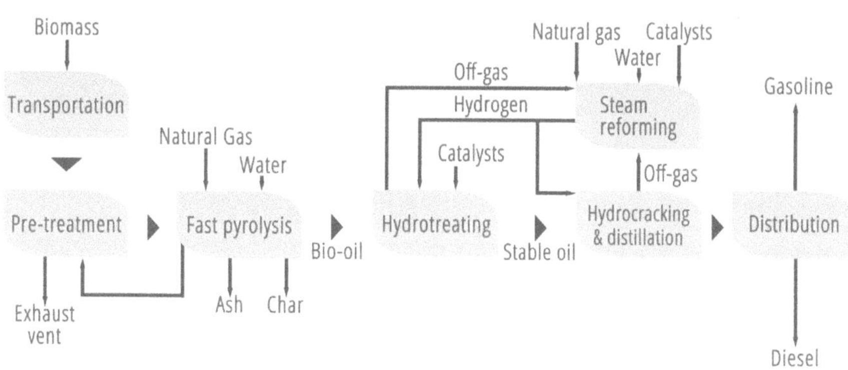

Figure 5.5 Conversion of biomass to biofuels via pyrolysis: value-chain process steps (Iribarren et al., 2012).

There are several existing pyrolysis reactor technologies that, among others, differ on the heat transfer mode to the biomass feedstock (Isahak et al., 2012). The source of the provided heat plays an important role in the subsequent LCA section. Using fossil-based sources as heat providers would, of course, deteriorate the environmental performance of the process. A proposal is to use a part or even the total amount of the produced char and permanent gases to support the endothermicity and increase the overall efficiency of the process (Bridgwater, 2012).

The third section includes the separation of the three products. The first step is to rapidly remove the produced biochar from the pyrolysis reactor to avoid cracking and other secondary reactions of the gases. This is usually accomplished by cyclones and hot filters (Bridgwater, 2012; Hoekstra et al., 2009; Mohan et al., 2006). Afterward, the vapor phase is rapidly quenched to form the oil and the permanent/uncondenced gaseous phase . The properties and characteristics of each product phase are described below:

- Bio-oil is the dark brown liquid product of biomass pyrolysis obtained at 60–70 wt.% of the biomass feed (Isahak et al., 2012). Pyrolysis bio-oils are complex organic mixtures containing more than 300 identified oxygenated compounds. Their composition and the intense presence of oxygen render bio-oils as highly acidic and viscous, thermally unstable, and not miscible with hydrocarbons (Bridgwater, 2012; Zhang et al., 2007). Bio-oils are investigated for usage as transportation fuels or as potential candidates for insertion into existing refinery systems (Talmadge et al., 2014; Huber and Corma, 2007; van Dyk et al., 2019) but require further upgrading to reduce their oxygen and moisture contents. Bio-oils can be upgraded through a variety of physical or chemical upgrading techniques, including hydrodeoxygenation, catalytic vapor cracking, emulsification, or distillation. Other potential bio-oil applications are as combustion fuels or for the production of chemicals (Czernik and Bridgwater, 2004).
- Biochar is the solid carbonaceous pyrolysis product obtained in yields around 12–15 wt.% (Isahak et al., 2012). Despite their low mass yield, it contains a significant portion of the energy content of the biomass feed. It can be used for power and heat production in combustion processes and is able to provide the necessary heat for the pyrolysis process. It could also act as a vapor cracking catalyst, and therefore, the rapid removal of the formed char should be performed to avoid secondary cracking reactions (Bridgwater, 2012).

- Permanent gases refer to the gases that remain uncondensed after the quenching stage and are obtained in yields around 13%–25% (Isahak et al., 2012). The composition and the energy content of the gas depend, among others, on the pyrolysis temperature (Ningbo et al., 2015), the pyrolysis configuration (usage of inert gas during the reaction) (Encinar et al., 2000), and condensation temperature (Choi et al., 2012). The heating value of the gases is not sufficient to provide the necessary pyrolysis heat and should be mixed/enriched with a higher-value gas for this purpose (Bridgwater, 2012).

Global warming potential (GWP) in each step of biofuel production via fast pyrolysis of corn stover and hydroprocessing is presented in Fig. 5.6.

Bio-oil upgrading is the largest contributor to total GWP, driven primarily by the consumption of conventional natural gas and energy for hydrogen production. The pyrolysis process is the second-largest contributor, resulted from large proportions of energy use. However, the heat recovery from coproduct biochar in pyrolysis plants can make pyrolysis attractive in the whole conversion pathway. Biofuel distribution and biomass transportation processes show the most negligible contribution individually in this conversion pathway, while biomass pretreatment

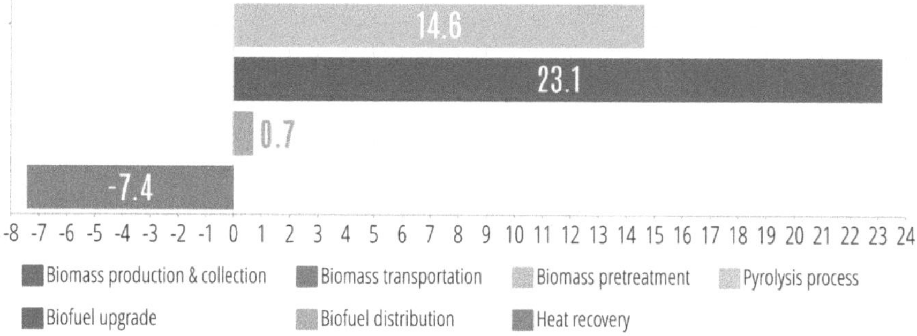

Figure 5.6 GWP along the life cycle of the production of 1 MJ biofuel from corn stover (Dang et al., 2014).

consumes enormous amounts of energy for the evaporation of moisture content.

5.3.2 Conversion of biomass to syngas via gasification

Biomass gasification denotes the thermochemical conversion of solid biomass feedstock primarily into synthesis gas, also known as "syngas"—predominately consisted of CO and H_2 (Shahabuddin et al., 2020). Despite that syngas is produced, by most types of optimized gasification technologies, at high temperatures and under well-controlled conditions, it is expected to contain significant amounts of tars and ash as contaminants or by-products (dos Santos and Alencar, 2020).

In principle, biomass gasification occurs via a set of sequential and complex reactions, which divide the gasification reactor ("gasifier") into four zones: (1) drying or bunker zone, (2) pyrolysis or thermal decomposition zone, (3) partial oxidation or combustion zone, and (4) reduction zone. The majority of gasification technologies are completely autothermal and require oxygen or another oxidizing agent like air, steam, steam/oxygen mixture, etc. (Sansaniwal et al., 2017). Some technologies may require minimal or no biomass treatment, while biomass pretreatment is a prerequisite for others. Biomass drying, particle size reduction, and purification from contaminants are the common pretreatment steps, while mechanical and/or biological treatment can also be applied to facilitate biomass gasification (Shahabuddin et al., 2020).

Generally, thermal gasification technologies are categorized into three basic classes: (1) fixed (moving) bed, that is, downdraft or cocurrent, updraft or counter-current, and cross-draft, (2) fluidized bed, that is, bubbling bed and circulating bed, and (3) entrained flow gasifiers (Ren et al., 2020). More recent technologies, such as catalytic, plasma, and supercritical H_2O, and chemical looping gasifiers, present significant advantages over conventional thermal gasification, including more efficient carbon conversion, lower emissions, reduced formation of inert ash and tars, and a better quality of gas-fuel products, though with increased associated cost (Farzad et al., 2016).

The design and operation of a biomass gasifier strongly depend on the feedstock availability, its properties, and the targeted end-use applications (Sansaniwal et al., 2017). The major parameters affecting the process and the quantity and quality of products are related to the biomass feedstock (e.g., biomass type and composition, moisture content, particle size, and purity) and the process conditions (e.g., bed material, gasifying agent type, and pressure, heating rate and temperature, gasification pressure, gasifying agent to biomass ratio, and residence time) (Molino et al., 2018).

As gasification converts biomass primarily into not only gaseous products but also liquids (tars) and solid residues (ash), the anticipating product's composition is crucial for evaluating the end-use applications (Brown, 2019). The most important product of biomass gasification is syngas, a well-known source of valuable chemicals, that is, diesel or gasoline (via Fischer-Tropsch synthesis), hydrogen, fertilizers, methanol, etc. It can also be used directly in electricity generators or boilers to produce heat and/or energy. Syngas from a standard biomass gasifier contains H_2 and CO as major components and CO_2, CH_4, H_2O, trace amounts of higher hydrocarbons (HCs), possible inert gases present in the gasification agent, and various contaminants. Syngas yield and composition depend on the feedstock properties, the operating conditions, and the gasifying agent type; oxygen is the most appropriate gasifying agent for syngas production (Akhtar et al., 2018). Further, biomass gasification can lead to the efficient production of hydrogen under well-defined conditions. High hydrogen yield can be obtained through various reforming—over chemical catalysts—technologies, such as steam, dry or autothermal reforming, and partial oxidation (Pandey et al., 2019).

The contaminants and by-products in biomass-derived syngas are usually managed separately. Syngas carries particulates and light HCs that should be cleaned up before its final use or conversion. Thus, syngas cleanup generally constitutes a critical and costly procedure involving an integrated, multistep approach. Gas-phase impurities include NH_3, HCN, other N-containing gases, H_2S, other S-containing gases, HCl, alkali metals, organic HCs (including tars), and particulates (Abdoulmoumine et al., 2015). On the other hand, tars contain the condensable organic fraction of gasification products. Several "in-line" and "on-line" tars sampling protocols have already been developed in order to standardize the way they are recovered. State-of-the-art technologies for syngas cleanup and tars mitigation include gas cyclones to remove most particulate matter, barrier filters, electrostatic precipitators, or liquid scrubbers/absorbers for the remaining particulate matter. It is also possible to catalytically crack tars into noncondensable gases at high temperatures via thermal oxidation and thermal/catalytic reforming (Brown, 2019).

Overall, the large-scale application of present biomass gasification technologies, a value chain consisting of several processing steps, such as biomass pretreatment, gasification, downstream syngas cleanup, and utilization (Fig. 5.7), is confined by several technical, economic, and environmental aspects.

Very recently, high-end advanced technologies have been proposed as the means to significantly improve the sustainability of the biomass gasification value chain (Dou et al., 2019). In more detail, gasification and cogeneration are approaches to combine the production of two or more products and to maximize the feedstock conversion efficiency. In addition, modern gasification concepts separate pyrolysis from the remaining steps in a single controlled stage that produces syngas with high purity (low tars levels). Finally, recent developments in innovative catalysts, sorbents, and high-temperature filtration offer the opportunity to integrate biomass gasification and gas cleanup in a single unit (Farzad et al., 2016). Fig. 5.8 shows the GWP that is related to the production and use of 1 m^3 of syngas.

GWP potential of biomass production and collection are related to the use of fossil fuels. Biomass transport by truck over short distances is rather efficient, and thus, the use of diesel and generated air emissions only cause small impacts. Syngas production is the second most contributing stage, related to NO_x and particulate emissions from gasification and cogeneration in the gas engine. Natural gas that is used for the start-up of the process also contributes to GWP. Syngas distribution presents major impacts if local networks are used (in the case of syngas for heating and Combined Heat and Power - CHP), which exhibit significantly higher methane leakages. GHG emissions are also

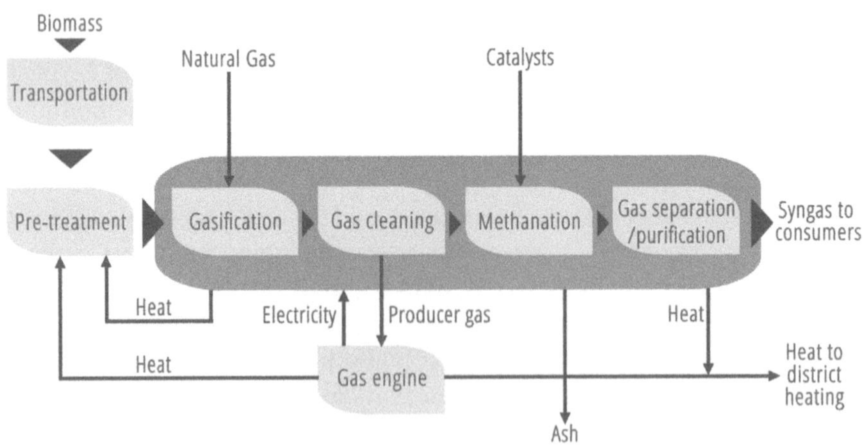

Figure 5.7 Conversion of biomass to syngas via gasification: value-chain process steps (Steubing et al., 2011).

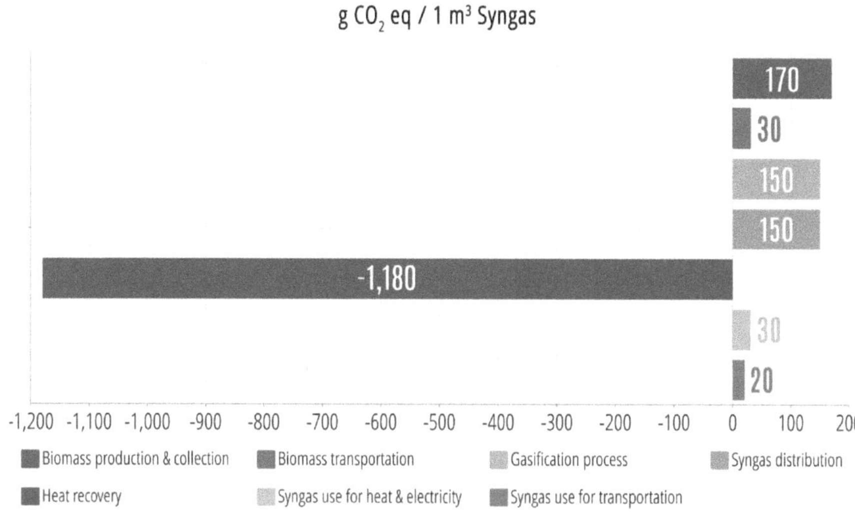

Figure 5.8 GWP along the life cycle of the production and use of 1 m^3 syngas from wood residues (Steubing et al., 2011).

generated during the use of syngas, for example, due to the electricity needed for boiler operation, tyre and brake pad abrasion that lead to soil and groundwater contamination with heavy metals, etc. Considerable impact reductions can be achieved from the use of process heat to replace heat from natural gas.

5.3.3 Overview of selected studies

Several LCA studies have been conducted with the objective of examining the environmental impact of pyrolysis and gasification systems. This section describes their scope and presents an overview of the considered biomass feedstock, the intended application of end product, and methodological choices made in preparation of the LCA studies. Feedstock source in biomass utilization LCA studies is a very important aspect because it is related to concerns of competition with food availability, actual net energy output, consequences of land-use change, the addition of agrochemical input, the effects of removing biomass residues from soil, and energy efficiency of biomass pretreatment. Feedstock classification affects the way the environmental burden is attributed (Wiloso et al., 2012) (Table 5.1).

The end product intended application refers to the fulfillment of a specific need for which the product was produced. It is crucial for "cradle-to-grave or well-to-wheel" LCAs that the end use is clearly stated since it directly

Table 5.1 Types of FU and their relevant research questions.

Type	Example of FU	Example of research question	Purpose
Feedstock related	1 kg biomass/ 1 MJ of biomass	What technological pathway is best for the conversion of this biomass?	Comparing the best uses for a given biomass feedstock
Agricultural land-related	1 ha of land	What is the best use of land?	Determining the best use of land
Single product related	1 kg of product /1 m^3 of product/ 1 MJ of product	What is the environmental impact of the production process?	Compare different technologies
Function of a single product related	1 MJ of electricity-heat/ 1 km of transportation	How is the environment affected by the use of the product?	
Multiple products related	1 process unit/ combination of output products	What are the hotspots in the production system? What is the optimal combination of output products for reducing environmental impact?	Identifying hotspots, comparing supply of the same functions based on fossil fuels, comparing stand-alone plants with integrated systems

impacts their environmental performance and affects the basis of comparison. Finally, as already stated, methodological choices strongly affect LCA results (Cherubini and Strømman, 2011). Tables 5.2 and 5.3 provide an indicative set of LCA studies in the field of thermochemical conversion technologies of biomass (gasification and pyrolysis, respectively).

5.3.3.1 *Feedstock*

The selected studies cover a wide range of biomass feedstocks. A distinction has to be made between biomass derived from dedicated crops

Table 5.2 List of LCA studies on gasification of biomass.

Gasification						
Feedstock	**Boundary**	**End product**	**FU**	**Multifunctionality**	**Impact categories**	**Study**
Wood	Cradle-to-grave	Heat	900,000 GJ/year delivered steam	NS	GWP, AP, POP, HTP,	Chan et al. (2008)
Forest residues	Cradle-to-grave	Electricity and heat	1 MJ of thermal energy–1 MJ of electricity	Exergy	GWP, AP, ADP, EP, POP, OLDP, HTP, FATP, MATP, TTP	Guest et al. (2011)
Forest residues	Cradle-to-grave	Electricity	1 kWh	Energy and Exergy	GWP, CED, AP, EP	Hennig and Gawor (2012)
Forest and wood residues	Cradle-to-grave	Electricity and heat	0,5 MW	System expansion	GWP	Cambero et al. (2015)
Forest and wood residues	Cradle-to-grave	Electricity and heat	2 MW	System expansion	GWP	
Forest and wood residues	Cradle-to-grave	Electricity and heat	5 MW	System expansion	GWP	
Energy crop (poplar)	Cradle-to-gate	H_2	1 kg H_2	Ignore	GWP, CED, OLDP, POP, AP, EP, LU	Susmozas et al. (2016)

(Continued)

Table 5.2 (Continued)

Gasification						
Feedstock	**Boundary**	**End product**	**FU**	**Multifunctionality**	**Impact categories**	**Study**
Energy crop (willow)	Cradle-to-grave	Electricity and heat	1-year supply of heat (including hot water) and electricity to a modern village of 150 households	Economic	GWP, FER, PER, LU, AP	Kimming et al. (2011)
Forest residues	Cradle-to-grave	Syngas	1 m^3 of syngas	System expansion	GWP, ADP, AP, EP, LU, POP, RINP, CED, Ecoindicator 99, Ecological Scarcity 2006	Steubing et al. (2011)
Forest wood	Cradle-to-gate	H_2	1 kg H_2	Energy	GWP, EU	Salkuyeh et al. (2018)
Agricultural residues	Gate-to-gate	Electricity	1 kWh	NS	HTP, FATP	Chidikofan et al. (2017)
Biowaste	Cradle-to-grave	Biomethane	1 MW	System expansion	GWP, RINP, NRE	Ardolino and Arena (2019)
Biomass	Cradle-to-gate	H_2	1 MJ	NS	Ecoindicator 95	Koroneos et al. (2008)

Energy crop (willow)	Cradle-to-grave	Electricity	Quantity of chips from 1 ha	Avoided	GWP, ADP, AP, EP, LU, OLDP, POP, CED	González-García et al. (2012)
Waste wood	Cradle-to-gate	Syngas	1 MJ of syngas	Energy	GWP, Ecoindicator 99, Ecological Scarcity 2006	Luterbacher et al. (2009)
Wood residues	Cradle-to-grave	Heat	1 tn wood residues	Mass	HH, EQ, CC, R, PE (endpoint)	Pa et al. (2011)
Forest wood	Cradle-to-gate	Electricity and heat	1 kWh	Exergy	GWP, ADP, HTP, TTP, PMFP	Oreggioni et al. (2017)
Energy Crop	Cradle-to-grave	Electricity	1 MWh	System expansion	GWP	Thornley et al. (2015)
Forest residues	Cradle-to-grave	Ammonia	1 tn	System expansion	GWP	Thornley et al. (2015)
Vineyard pruning	Cradle-to-grave	Electricity	1 MWh	Economic, system expansion	GWP, AP, ADP, OLDP, EP, HTP, TTP, FATP, MATP, POP, CED	Hamedani et al. (2019)
Agricultural residues	Cradle-to-grave	Electricity and heat	1 kWh	Energy	GWP, AP, EP, NRER, RINP	Nguyen et al. (2013)

(Continued)

Table 5.2 (Continued)

Gasification

Feedstock	Boundary	End product	FU	Multifunctionality	Impact categories	Study
Waste wood	Cradle-to-gate	Electricity and heat	1000 MJ of primary energy	Energy	GWP, ADP, OLDP, HTP, AP, EP	Puy et al. (2010)
Waste wood	Cradle-to-gate	Electricity and heat	1000 MJ of primary energy	Energy	GWP, ADP, OLDP, HTP, AP, EP	
Forest Residues	Cradle-to-gate	Electricity and heat	1000 MJ of primary energy	Energy	GWP, ADP, OLDP, HTP, AP, EP	

Table 5.3 List of LCA studies on pyrolysis of biomass.

Pyrolysis						
Feedstock	**Boundary**	**End product**	**FU**	**Multifunctionality**	**Impact categories**	**Study**
Forest residues	Cradle-to-grave	Electricity	1 kWh	Energy	GWP, AP, EP, POP, OLDP,	Faix et al. (2010)
Energy crops (willow, poplar), forest residues, waste wood	Cradle-to-grave	Electricity	1 kWh	NS	GWP	Fan et al. (2011)
Forest wood	Cradle-to-grave	Bio-oil	1 MJ	System expansion and Energy	GWP, CED	Winjobi et al. (2016)
Forest residues	Cradle-to-grave	Diesel and Gasoline	1 km	Energy	GWP, NEV	Hsu (2012)
Agricultural residues	Cradle-to-gate	Bio-gasoline	1 ha	NS	GWP	Kauffman et al. (2011)
Energy crop (poplar)	Cradle-to-gate	Gasoline, diesel and char	43478 MJ	Economic	GWP, OLDP, POP, LU AP, EP, CED	Iribarren et al. (2012)
Agricultural residues	Cradle-to-grave	Gasoline	1 MJ, 1 tn agricultural residues	Energy and System expansion	GWP	Han et al. (2013)

(Continued)

Table 5.3 (Continued)

Pyrolysis						
Feedstock	**Boundary**	**End product**	**FU**	**Multifunctionality**	**Impact categories**	**Study**
Energy crop (willow)	Cradle-to-gate	Biomethane	1 ha, 1 GJ	Energy and System expansion	GWP, PER	Moghaddam et al. (2019)
Agricultural residues	Cradle-to-grave	Biochar	1 tn residues	System expansion	GWP, NER	Roberts et al. (2010)
Yard waste	Cradle-to-grave	Biochar	1 tn waste	System expansion	GWP, NER	
Energy crop (switchgrass)	Cradle-to-grave	Biochar	1 tn switchgrass	System expansion	GWP, NER	
Agricultural residues	Cradle-to-gate	Bio-oil	1 kg bio-oil	NS	GWP, AP, EP, HTP, POP	Chan et al. (2016)
Energy crop	Cradle-to-grave	Biochar	1 tn	System expansion	GWP	Thornley et al. (2015)
Agricultural residues	Cradle-to-grave	Electricity and heat, biochar	1 tn of corn stover	Energy	GWP, AP, EP, ADP, PED	Righi et al. (2016)
Energy crop (poplar)	Cradle-to-grave	Heat	1 ha	System expansion	GWP, ADP, AP, EP, CED	Peters et al. (2015a)

Energy crop (poplar)	Cradle-to-grave	Biochar	1 ha	System expansion	GWP, ADP, AP, EP, CED	
Energy crop (poplar)	Cradle-to-grave	Biochar	1 ha	System expansion	GWP, ADP, AP, EP, CED	
Agricultural residues	Cradle-to-gate	Bio-oil	1 kg of corn stover	Mass	GWP, AP, ADP, OLDP, HTP, EP	Han et al. (2019)
Agricultural residues	Cradle-to-gate	Biofuel	1 kg of upgraded fuel	System expansion	GWP, AP, EP, OLDP, ADP, FATP, HTP, MATP, POP, TTP	Vienescu et al. (2018)
Forest wood	Cradle-to-grave	Diesel	1 MJ	Energy	GWP, NER	Wong et al. (2016)
Forest residues	Cradle-to-grave	Diesel	1 MJ	Energy	GWP, NER	
Agricultural residues	Cradle-to-grave	Diesel	1 MJ	Energy	GWP, NER	
Forest residues	Cradle-to-grave	Electricity	1 kWh	NS	GWP	Gu and Bergman (2015)

(Continued)

Table 5.3 (Continued)

Pyrolysis						
Feedstock	**Boundary**	**End product**	**FU**	**Multifunctionality**	**Impact categories**	**Study**
Forest residues	Cradle-to-gate	Bio-oil, biochar	1 gallon bio-oil	NS	GWP	Mirkouei et al. (2017)
Forest residues	Cradle-to-grave	Bio-oil, biochar	1 tn of green thinned logs	Economic	GWP, AP, EP, FDP, HTP	Lu, and El Hanandeh (2019)
Energy crop (miscanthus)	Cradle-to-gate	Bio-hydro-carbons	1 tn of bio-hydro-carbons	Energy	GWP	Shemfe et al. (2016)

GWP, Global warming potential; *AP*, acidification potential; *EP*, Eutrophication potential; *ADP*, abiotic depletion potential; *LU*, land use; *POP*, photochemical oxidation potential; *OLDP*, ozone layer depletion potential; *HTP*, human toxicity potential; *FATP*, freshwater aquatic toxicity potential; *MATP*, marine aquatic toxicity potential; *TTP*, terrestrial toxicity potential; *CED*, cumulative energy demand; *RINP*, raspatory inorganics potential; *NER*, net energy ratio; *PED*, primary energy demand; *NRE*, nonrenewable energy; *EU*, energy use; *PER*, primary energy requirement; *FER*, fossil energy requirement; *NEV*, net energy value.

(Susmozas et al., 2016; Kimming et al., 2011; etc.) and the residues of agricultural (Chidikofan et al., 2017; Han et al., 2019; etc.) and forestry (Cambero et al., 2015; Wong et al., 2016; etc.) operations. Energy crops tend to create competition with fertile land and food and feed crops, but on the other hand, their energy content is immense. On the other side, agricultural residues are an underutilized low-cost feedstock source for bioenergy applications, with, however, a huge contribution to the maintenance of soil organic carbon stocks.

An excellent example of energy crops versus agricultural residues is presented in the study by Thornley et al. (2015). The methodology adopted for this work involved completing a full LCA for each of six different bioenergy systems. The scope of the system encompassed feedstock production (ground preparation, establishment, harvesting, and restoration), processing (transport, drying, storage, loading, etc.), and conversion to the final product (electricity, chemical product, heat, etc.). The comparison between woodchips from energy crops for small-scale electricity applications and woodchips from imported forest residues for large-scale electricity applications showed that the GHG savings from the bioenergy system per unit of product (1 MWh) are almost the same (less than 1% higher savings from large-scale applications). However, the GHG reductions per unit of land occupied (ha per annum) revealed that the savings from the use of forest residues are 65% higher than the savings from the use of energy crops. These results are increasing the concerns about the competition between food and fuel and the awareness of the environmental, social, and economic implications of LU for bioenergy (McDowall et al., 2012).

Wood is the most investigated biomass type, probably because it can be extracted from many sources: forestry, short rotation forestry, sawmills, waste wood, etc. Many studies also analyze agricultural residues like corn stover and wheat straw; palm oil residues are restricted to regions with suitable climate conditions (Chan et al., 2016). The studies with nonconventional sources of biomass were limited to biowaste (Ardolino and Arena, 2019) and yard waste (Roberts et al., 2010).

Wong et al. (2016) studied the GHG emissions of three feedstocks: whole tree (i.e., chips from cutting the whole tree), forest residues (i.e., chips from branches and tops generated from logging operations), and agricultural residues (i.e., straw from wheat and barley) for the production of hydrogenation-derived renewable diesel (HDRD). They concluded that the differences in results among feedstocks (35.4 gCO_2,equiv./MJ of

HDRD for agricultural residues, 39.7 gCO_2,equiv./MJ of HDRD from cutting of whole trees, and 42.3 gCO_2,equiv./MJ of HDRD for forest residues) are due to different energy requirements to harvest and pretreat biomass. Also, the energy-intensive hydroprocessing stage is responsible for most of the GHG emissions produced for the entire conversion pathway.

Ultimately, comparing feedstocks shows the significance of the biomass source, the efficiency in the equipment used for harvest, transportation, and distribution, the physical properties of biomass in the production of bioenergy, and the importance of efficiency at the biomass conversion stage.

5.3.3.2 Intended application of end product

The majority of studies are focused on stationary applications and particularly on electricity and heat generation. Very few studies focused on soil application and chemicals (Thornley et al., 2015; Roberts et al., 2010; Righi et al., 2016; Peters et al., 2015a). In comparison to stationary applications, the number of studies evaluating the environmental performance of biomass for transportation biofuels is limited and can be attributed to the increasing popularity of electric, hybrid electric, and fuel cell electric vehicles. Koroneos et al. (2008), Susmozas et al. (2016), and Salkuyeh et al. (2018) studied hydrogen production via biomass gasification, but they did not explore the potential applications.

5.3.3.3 Scope of the selected studies

The FU has an integral part in any LCA, as it is the reference point of all relevant flows and impact assessment results. A FU in correlation with the output unit, either in energy, mass, or distance, is selected by studies comparing various feedstocks or comparing bioenergy with fossil-fuel product systems. The majority of the studies selected an output-related FU, even studies with feedstocks derived from energy crops (Susmozas et al., 2016; Kimming et al., 2011; Fan et al., 2011; Iribarren et al., 2012; Shemfe et al., 2016). This is very important since the results of a bioenergy system expressed per agricultural land unit indicate land competition and land-use efficiency. In that matter, very few studies selected a FU related to the agricultural land needed for the cultivation or extraction of the feedstock (González-García et al., 2012; Kauffman et al., 2011; Peters et al., 2015b). Quantity of feedstock as FU or input related was selected by studies comparing uses of a specific feedstock or attempting to avoid allocation

(Han et al., 2019; Roberts et al., 2010; Righi et al., 2016). Han et al. (2013) and Moghaddam et al. (2019) used more than one FU, one input and one output-related FU in the first case and one output and one agricultural land-related in the latter. This enabled a greater understanding of the product systems under investigation.

Only (Hsu, 2012) used an output-related FU expressed per vehicle-kilometer basis concerning studies on transportation fuels. This is problematic because only this kind of FU can ensure that all life cycle stages are considered, especially the mechanical efficiency of the transportation fuel and the comparability with conventional fossil systems.

The system boundary determines the activities or processes included and assessed and those excluded from the analysis. The definition of system boundary as cradle-to-gate or cradle-to-grave is very broad and sometimes insufficient for thermochemical conversion routes as the omission of life cycle stages is allowed if it does not significantly affect the outcome of the LCA. An example is provided by two studies in which it was stated that a cradle-to-grave system boundary was chosen; on the one hand, Guest et al. (2011) considered the distribution of electricity, and on the other hand, Hennig and Gawor (2012) set it outside the scope of the study. This might not affect the results, but their comparison is impossible due to this discrepancy. In regard to biomass life cycle, the cultivation was included in all the studies concerning energy crops and excluded from those that utilize residues. Biomass collection, preparation, and transportation were included in all the studies except Chidikofan et al. (2017), which is justified by the gate-to-gate approach. In general, biomass logistics is a very important aspect of bioenergy. Even if the purpose of an LCA study is to compare conversion technologies, feedstock-related issues should always be addressed. In regard to the conversion system, the capital equipment is not considered within the scope of most of the studies, as it is suggested to have a negligible environmental impact on LCA results. Concerning end-of-life treatment, very few studies included this stage and only those that consider the application of char as a soil amendment or as a substitute for fossil fuels.

5.3.3.4 Multifunctionality

Handling multifunctionality is an important issue for thermochemical conversion systems because they usually are characterized by multiple useful products that share the total environmental impact. According to the ISO, the allocation should be avoided whenever possible by system expansion. The environmental impact is calculated as the emissions from the main product minus the emissions from the use of the coproducts. Another way of avoiding allocation is by choice of FU, usually an input-related like the quantity of feedstock. When system expansion is not possible, or the goal of the study dictates another type of FU, different allocation methods should be applied based on physical relationships or economic value.

Most studies expand the system then apply the substitution method, while others share the environmental impact among various products based on the energetic content. This approach sometimes causes problems because it does not take into account the quality of the form of energy. Guest et al. (2011), Hennig and Gawor (2012), and Oreggioni et al. (2017) used exergy as the basis of allocation. Exergy takes into account the thermodynamic significance of every energy form. Economic-based allocation was used by Lu and El Hanandeh (2019), Iribarren et al. (2012), Hamedani et al. (2019), and Kimming et al. (2011). Although it is subject to regional and temporal price differences, Lu and El Hanandeh (2019) argued that economic allocation is recommended when the main product and coproduct values diverge greatly. Mass-based allocation is not common and is applied only by Pa et al. (2011) and Han et al. (2019). A few studies avoided allocation by allocating all the environmental burdens to the main product. A very common issue is that many studies do not explicitly specify how environmental impacts are allocated between coproducts, thus making the interpretation of results meaningless.

5.3.3.5 Life cycle impact assessment methods

The majority of LCA studies on the thermochemical processing of biomass employed midpoint categories, whereas the endpoint approach was used only by Pa et al. (2011). Hybrid approach (combination of midpoint and endpoint categories) was utilized by Steubing et al. (2011) and Luterbacher et al. (2009). The characterization method applied was not explicitly stated in most of the studies, but the reported impact indicators use units equivalent to those of the CML methodology.

GWP was the most addressed category, followed distantly by acidification potential and eutrophication potential. Another aspect that widely received attention was energy evaluation. Besides cumulative energy demand, many other different energy indicators were applied, like net energy ratio, energy use, primary energy requirement, fossil energy requirement, net energy value, etc., that in most cases are not comparable.

A limited number of studies (Susmozas et al., 2016; Kimming et al., 2011; González-García et al., 2012; Iribarren et al., 2012; Steubing et al., 2011), assessed the impact of LU change on the overall environmental footprint of the thermochemical conversion systems. This indicator is crucial since LU change affects biodiversity and soil quality, especially in systems based on energy crops and residues; however, the absence of a widely accepted LU change evaluation methodology partially justifies the exclusion. The impact on human health and water are also important issues that are not well covered by the reviewed studies by Guest et al. (2011), Chidikofan et al. (2017), Vienescu et al. (2018), and Hamedani et al. (2019).

5.4 Issues affecting the comparability of life cycle assessment studies

LCA is a well-known and established method for evaluating environmental impacts. However, LCA of thermochemical conversion routes still faces issues regarding the basic methodological choices in LCA, for example, choice of FU, handling multifunctionality, source of data, system boundaries, and LCIA method. The main objective of this section is to identify and discuss key methodological issues for LCA of pyrolysis and gasification in relation to selected literature. The intention is to uncover the complexities of such systems' LCA, which can be useful for LCA practitioners. Some of the issues are not specific to the LCA of thermochemical conversion but can generally be applied to LCA.

The literature review revealed major discrepancies in methodological choices made between selected studies. The following key issues affecting their comparability were identified:

- *Boundaries.* The system boundary determines which processes of a system are assessed. The definitions "cradle-to-grave" or "cradle-to-gate" are very broad and can lead to different assumptions even among studies with seemingly similar boundaries.
- *Functional unit.* Although in most cases the selection and formulation of the FU are very well defined and in line with the aim of the study, the comparison of the studies is impossible because they describe different functions, qualitatively and quantitatively.

- *Multifunctionality*. The different ways of handling multifunctionality cause the systems compared not to encompass the same multiple functions.
- *Source of data*. The plethora of sources for data collection is a blessing and a curse. On the one hand, the data collection process is greatly simplified, and the time required is reduced. On the other hand, both the uncertainty regarding the quality of the data and the differentiation of the assumptions made for the calculation and/or collection of this data increase.
- *Impact categories*. Despite the fact that a vast number of environmental impact categories exist, a common element that all studies possess (except one) is the GWP calculation and can be considered a common ground for comparison. However, the problem lies in the characterization factors utilized by each LCIA methodology for the formulation of the category indicator result.

All of these large or small inconsistencies make comparisons between studies complex or near impossible.

Regardless of the fact that there are many LCA standards and guidelines to provide specific methodological recommendations relevant to LCA of thermochemical conversion routes, a stricter harmonization of LCA studies is necessary to improve their comparability. To this end, the use of key performance indicators (KPIs) and/or the adoption of more strict, robust methods for evaluating life cycle impacts can significantly affect the way LCA is practiced in the future.

5.4.1 Key performance indicators

Indicators are quantitative or qualitative variables providing means to help assess performance or compliance in the areas of concern, reflect changes, and measure achievement (Panoutsou et al., 2020). The KPIs are sets of metrics reflecting performance for resource efficiency, economic, environmental, and social sustainability (Yu et al., 2018). They could be extremely useful if combined with LCA methodology to enable the comparability of different products and technologies across their life cycle. KPIs must comply with three main features (Palomba and Frazzica, 2019):

- simplicity, in order to be understood not only by technicians or researchers specialized in the field but also by stakeholders of interconnected sectors

- clear and unique definition, in order not to misunderstand or evaluate the parameter wrongly when applying it to different technologies
- meaningfulness, which means that the selected parameter should be of relevance not only to one stakeholder (e.g., the academia or the designer) but to several entities involved in the design/realization/operation/evaluation process.

Benefits from this combination are the identification of opportunities to improve the environmental performance of products and technologies, the adoption of new technologies, and the straightforward interpretation by the end users due to quantitative results (Dorn et al., 2016).

5.4.1.1 Key performances indicators and life cycle assessment of biomass thermochemical conversion routes

KPIs are related to the collection of all necessary information about inputs and outputs of a process during the second phase of an LCA (LCI-inventory) and the subsequent evaluation of the sustainability performance by life cycle aspects during the third phase (LCIA-impact assessment).

The review of the selected studies revealed that KPIs are overlooked in LCAs of biomass thermochemical conversion. This could be due to the complex interrelations of physical assets (land, soil, water, air, and climate) with market attributes (displacement of other land-based activities, competition for raw materials, alternative technologies, and valorization of coproducts) within the individual stages of biomass value chains (Lokesh et al., 2018) that hamper the formulation of relevant KPIs. However, from the analysis of the individual life cycle stages of thermochemical conversion routes, it is clear that a set of indicators that can simplify the measuring and comparison should at least include

- Energy used from fossil fuels
- Land efficiency
- Life cycle GHG emissions
- Air quality emissions
- Life cycle costs
- Employment along the life cycle
- Contribution to the rural economy

An indicative set of suggested KPIs for life cycle stages and activities related to thermochemical conversion pathways are presented in Table 5.4. The formulation of the KPIs was based on the following principles

Table 5.4 Indicative set of KPIs to measure, monitor, and interpret the performance of each life cycle stage of biomass thermochemical conversion pathways.

Life cycle stage				
		Biomass production	**Conversion to bioenergy carriers**	**End-use (transportation, energy commodity)**
KPIs	Resource efficiency	Annual biomass production per unit of cultivated area (tn/ha/year)		
		Product to by-product ratio (tn/tn biomass)	Product to by-product ratio (tn or GJ/GJ end-product)	
		Annual end-product per unit of cultivated area (GJ/ha/year)		
			End-product per mass of feedstock (tn/tn or GJ/tn)	
		Cumulative energy demand (GJ input/GJ end-product)	Cumulative energy demand (GJ input/GJ end-product)	Cumulative energy demand (GJ input/GJ end-product)
		Renewable energy demand (GJ input/GJ end-product)	Renewable energy demand (GJ input/GJ end-product)	Renewable energy demand (GJ input/GJ end-product)
		Nonrenewable energy demand (GJ input/GJ end-product)	Nonrenewable energy demand (GJ input/GJ end-product)	Nonrenewable energy demand (GJ input/GJ end-product)
				Efficiency (km or GJ output/GJ input)
	Environment	GHG emissions (gCO_2equiv./GJ end-product)	GHG emissions (gCO_2equiv./GJ end-product)	GHG emissions (gCO_2equiv./GJ end-product)
		Sustainable harvest level (% of net annual growth)		
		Soil carbon and nutrients (tn/ha GJ of end-product)		

	Environment	Acidification (g SO_2 equiv./GJ end-product) Ozone formation (g NO_x equiv./GJ end-product) Particulate matter (g PM10/GJ end-product) Water use (m^3/GJ end-product) Wastes (tn/GJ of end-product)	Acidification (g SO_2 equiv./GJ end-product) Ozone formation (g NO_x equiv./GJ end-product) Particulate matter (g PM10/GJ end-product) Water use (m^3/GJ end-product) Ash (tn/GJ end-product) Wastes (tn/GJ of end-product)	Acidification (g SO_2 equiv./GJ end-product) Ozone formation (g NO_x equiv./GJ end-product) Particulate matter (g PM10/GJ end-product) Ash (tn/GJ end-product) Wastes (tn/GJ of end-product)
	Economy	CAPEX (€/GJ end-product) OPEX (€/GJ end-product) Maintenance and repair (€/GJ end-product)	CAPEX(€/GJ end-product) OPEX (€/GJ end-product) Maintenance and repair (€/GJ end-product)	CAPEX (€/GJ end-product) OPEX (€/GJ end-product) Maintenance and repair (€/GJ end-product)
	Society	Number of full-time jobs/GJ of end-product Contribution to rural economy (€/GJ end-product)	Number of full-time jobs/GJ of end-product Contribution to rural economy (€/GJ end-product)	Number of full-time jobs/GJ of end-product Contribution to rural economy (€/GJ end-product)

- Appropriate to explain the performance of the life cycle stage.
- Easy to understand by all relevant stakeholders.
- Easy to be interpreted.
- Clear to be measured and replicated.
- Sensitive to human and natural activities.

The life cycle of thermochemical conversion of biomass can be divided into the following three stages:

- *Biomass production*: It includes the following activities: soil acquisition, soil management, harvesting, pretreatment, storage, and transport.
- *Biomass thermochemical conversion pathways—conversion to bioenergy carriers*: The main activities are the construction and operation of the conversion installations.
- *End use of products*: Activities are related to distribution, use, and disposal.

5.4.2 Product environmental footprint assessment

The product environmental footprint (PEF) method is based on the life cycle approach. The main goal is to increase comparability between products of the same product category (and therefore also allow for comparisons and comparative assertions), applying the "comparability over flexibility" approach, meaning that by reducing the flexibility of methodological choices, the comparability of products increases (Bach et al., 2018). The increased comparability was expected to be achieved by predefining specifications for certain methodological aspects based on value choices (e.g., modeling of the end-of-life phase) and thus reducing the flexibility for which ISO 14040/44 (Finkbeiner et al., 2006) (the internationally agreed-upon standard for LCA) is known for. Furthermore, by developing product category-specific rules for certain product categories, additional specifications are determined. A 3-year pilot phase was finalized in April 2018. The method is now going through a transition phase before a final decision will be taken on how to implement the method in existing legislation.

A core suite of analytical principles shall be adhered to produce reliable, reproducible, and verifiable PEF studies. These principles provide overarching guidance in the application of the PEF method. They shall be considered with respect to each phase of PEF studies, from the definition of goal and the scope, through data collection, impact assessment, reporting, and verification of study outcomes. Users of this method shall observe the following principles in conducting a PEF study (Manfredi et al., 2012):

- Relevance. All methods used and data collected for the purpose of quantifying the PEF shall be as relevant to the study as possible.
- Completeness. Quantification of the PEF shall include all environmentally relevant material/energy flows and other environmental interventions as required for adherence to the defined system boundary, the data requirements, and the impact assessment methods employed.
- Consistency. Strict conformity to this method shall be observed in all steps of the PEF study to ensure internal consistency and comparability.
- Accuracy. All reasonable effort shall be taken to reduce uncertainties in product system modeling and the reporting of results.
- Transparency. PEF information shall be disclosed in such a way as to provide intended users with the necessary basis for decision-making and for stakeholders to assess its robustness and reliability.

5.5 Conclusions

Thermochemical pathways present a significant potential to convert biomass into high-value bioenergy products. In this chapter, a review of several LCA studies on pyrolysis and gasification technologies was carried out to understand the issues arising in applying LCA methodology. Regarding the outcome of LCA, the evaluation of environmental performance is complicated; differentiations in feedstock, conversion pathways, intended application, and methodological choices lead to a wide range of results in terms of net energy balances and net GHG emissions. Concerning methodological choices in the goal and scope definition, the cradle-to-grave approach has the advantage over cradle-to-gate. Gate-to-grave was utilized only in one study. This signifies the importance of biomass cultivation and feedstock type on the overall performance of bioenergy systems. However, boundary definitions are very broad, the inclusion or exclusion of functions must be adequately and thoroughly justified. According to LCA standards, the choice and formulation of a FU must be consistent with the goal and scope of the study. Adopting a FU must ensure that all the life cycle stages defined by the system boundaries are included. Energy crop-based conversion systems must also include an agricultural land-related FU to measure land-use efficiency. In the literature review, there is no uniformity concerning the FU chosen by the different analysts, thus making the LCA studies challenging to compare. Thermochemical conversion routes of biomass are characterized as multifunctional systems (multiple end products, e.g., gaseous, liquids, solids, and intended applications, e.g., electricity, heat, transportation fuels). The method for handling multifunctionality has a strong influence on the overall LCA results. System expansion is the preferable method, followed by allocation based on energy, economic values, exergy, and mass. Nevertheless, a lot of LCA studies do not specify how environmental impacts are allocated between functions.

Regarding methodological choices in the LCI stage, information on data sources is usually reported, high usage of scientific literature and LCA databases was found for both foreground and background processes. However, regional relevance, data age, and assumptions affect the quality and, therefore, the overall result.

Concerning methodological choices in the LCIA stage, most studies address the GWP, followed by acidification potential and eutrophication potential. Energy evaluation is also included, but the wide range of indicators hinders comparability. Land-use change is not widely studied in the LCA of the thermochemical processing of biomass mainly due to the lack of an established and widely accepted method. KPIs could be combined with LCA to produce consistent and comparable studies. Finally, a harmonized and robust method is required for the comparison of end products and alternative conversion pathways.

References

Abdoulmoumine, N., Adhikari, S., Kulkarni, A., Chattanathan, S., 2015. A review on biomass gasification syngas cleanup. Applied Energy 155, 294–307.

Ahlgren, S., Björklund, A., Ekman, A., Karlsson, H., Berlin, J., Börjesson, P., et al., 2015. Review of methodological choices in LCA of biorefinery systems-key issues and recommendations. Biofuels, Bioproducts and Biorefining 9 (5), 606–619.

Akhtar, J., Amin, N.S., 2012. A review on operating parameters for optimum liquid oil yield in biomass pyrolysis. Renewable and Sustainable Energy Reviews 16 (7), 5101–5109.

Akhtar, A., Krepl, V., Ivanova, T., 2018. A combined overview of combustion, pyrolysis, and gasification of biomass. Energy & Fuels 32 (7), 7294–7318.

Ardolino, F., Arena, U., 2019. Biowaste-to-biomethane: an LCA study on biogas and syngas roads. Waste Management 87, 441–453.

Bach, V., Lehmann, A., Görmer, M., Finkbeiner, M., 2018. Product environmental footprint (PEF) pilot phase—comparability over flexibility? Sustainability 10 (8), 2898.

Beis, S.H., Onay, Ö., Koçkar, Ö.M., 2002. Fixed-bed pyrolysis of safflower seed: influence of pyrolysis parameters on product yields and compositions. Renewable Energy 26 (1), 21–32.

Bridgwater, A.V., 2012. Review of fast pyrolysis of biomass and product upgrading. Biomass and Bioenergy 38, 68–94.
Brown, R.C. (Ed.), 2019. Thermochemical Processing of Biomass: Conversion into Fuels, Chemicals and Power. John Wiley & Sons.
Cambero, C., Alexandre, M.H., Sowlati, T., 2015. Life cycle greenhouse gas analysis of bioenergy generation alternatives using forest and wood residues in remote locations: a case study in British Columbia, Canada. Resources, Conservation and Recycling 105, 59–72.
Chan, B., Chan, B., Young, C., 2008. Wood Pellets for UBC Boilers Replacing Natural Gas Based on LCA. University of British Columbia, Vancouver.
Chan, Y.H., Tan, R.R., Yusup, S., Lam, H.L., Quitain, A.T., 2016. Comparative life cycle assessment (LCA) of bio-oil production from fast pyrolysis and hydrothermal liquefaction of oil palm empty fruit bunch (EFB). Clean Technologies and Environmental Policy 18 (6), 1759–1768.
Cherubini, F., Strømman, A.H., 2011. Life cycle assessment of bioenergy systems: state of the art and future challenges. Bioresource Technology 102 (2), 437–451.
Chidikofan, G., Benoist, A., Sawadogo, M., Volle, G., Valette, J., Coulibaly, Y., et al., 2017. Assessment of environmental impacts of tar releases from a biomass gasifier power plant for decentralized electricity generation. Energy Procedia 118, 158–163.
Choi, H.S., Choi, Y.S., Park, H.C., 2012. Fast pyrolysis characteristics of lignocellulosic biomass with varying reaction conditions. Renewable Energy 42, 131–135.
European Commission - Joint Research Centre - Institute for Environment and Sustainability: International Reference Life Cycle Data System (ILCD) Handbook - General guide for Life Cycle Assessment - Detailed guidance. First edition March 2010. EUR 24708 EN. Luxembourg. Publications Office of the European Union; 2010
Collard, F.X., Blin, J., 2014. A review on pyrolysis of biomass constituents: mechanisms and composition of the products obtained from the conversion of cellulose, hemicelluloses and lignin. Renewable and Sustainable Energy Reviews 38, 594–608.
Czernik, S., Bridgwater, A.V., 2004. Overview of applications of biomass fast pyrolysis oil. Energy & Fuels 18 (2), 590–598.
Dang, Q., Yu, C., Luo, Z., 2014. Environmental life cycle assessment of bio-fuel production via fast pyrolysis of corn stover and hydroprocessing. Fuel 131, 36–42.
Di Blasi, C., Signorelli, G., Di Russo, C., Rea, G., 1999. Product distribution from pyrolysis of wood and agricultural residues. Industrial & Engineering Chemistry Research 38 (6), 2216–2224.
Dorn, C., Behrend, R., Giannopoulos, D., Napolano, L., James, V., Herrmann, A., et al., 2016. A systematic LCA-enhanced KPI evaluation towards sustainable manufacturing in industrial decision-making processes. A case study in glass and ceramic frits production. Procedia Cirp 48, 158–163.
dos Santos, R.G., Alencar, A.C., 2020. Biomass-derived syngas production via gasification process and its catalytic conversion into fuels by Fischer Tropsch synthesis: a review. International Journal of Hydrogen Energy 45 (36), 18114–18132.
Dou, B., Zhang, H., Song, Y., Zhao, L., Jiang, B., He, M., et al., 2019. Hydrogen production from the thermochemical conversion of biomass: issues and challenges. Sustainable Energy & Fuels 3 (2), 314–342.
Encinar, J.M., Gonzalez, J.F., Gonzalez, J., 2000. Fixed-bed pyrolysis of Cynara cardunculus L. Product yields and compositions. Fuel Processing Technology 68 (3), 209–222.
Fahmy, T.Y., Fahmy, Y., Mobarak, F., El-Sakhawy, M., Abou-Zeid, R.E., 2020. Biomass pyrolysis: past, present, and future. Environment, Development and Sustainability 22 (1), 17–32.

Faix, A., Schweinle, J., Schöll, S., Becker, G., Meier, D., 2010. (GTI-tcbiomass) life-cycle assessment of the BTO®-process (biomass-to-oil) with combined heat and power generation. Environmental Progress & Sustainable Energy 29 (2), 193–202.

Fan, J., Kalnes, T.N., Alward, M., Klinger, J., Sadehvandi, A., Shonnard, D.R., 2011. Life cycle assessment of electricity generation using fast pyrolysis bio-oil. Renewable Energy 36 (2), 632–641.

Farzad, S., Mandegari, M.A., Görgens, J.F., 2016. A critical review on biomass gasification, co-gasification, and their environmental assessments. Biofuel Research Journal 3 (4), 483–495.

Finkbeiner, M., Inaba, A., Tan, R., Christiansen, K., & Klüppel, H. J. (2006). The new international standards for life cycle assessment: ISO 14040 and ISO 14044. *The international journal of life cycle assessment, 11*(2), 80-85.

González-García, S., Iribarren, D., Susmozas, A., Dufour, J., Murphy, R.J., 2012. Life cycle assessment of two alternative bioenergy systems involving Salix spp. biomass: bioethanol production and power generation. Applied Energy 95, 111–122.

Gu, H., Bergman, R., 2015. Life-cycle GHG emissions of electricity from syngas produced by pyrolyzing woody biomass. Proceedings of the 58th International Convention of Society of Wood Science and Technology. Jackson Lake Lodge, Grand Teton National Park, Wyoming, United States, pp. 376–389, June 7–12, 2015.

Guest, G., Bright, R.M., Cherubini, F., Michelsen, O., Strømman, A.H., 2011. Life cycle assessment of biomass-based combined heat and power plants: centralized vs decentralized deployment strategies. Journal of Industrial Ecology 15 (6), 908–921.

Hamedani, S.R., Del Zotto, L., Bocci, E., Colantoni, A., Villarini, M., 2019. Eco-efficiency assessment of bioelectricity production from Iranian vineyard biomass gasification. Biomass and Bioenergy 127, 105271.

Han, J., Elgowainy, A., Dunn, J.B., Wang, M.Q., 2013. Life cycle analysis of fuel production from fast pyrolysis of biomass. Bioresource Technology 133, 421–428.

Han, D., Yang, X., Li, R., Wu, Y., 2019. Environmental impact comparison of typical and resource-efficient biomass fast pyrolysis systems based on LCA and Aspen Plus simulation. Journal of Cleaner Production 231, 254–267.

Hennig, C., Gawor, M., 2012. Bioenergy production and use: comparative analysis of the economic and environmental effects. Energy Conversion and Management 63, 130–137.

Hoekstra, E., Hogendoorn, K.J., Wang, X., Westerhof, R.J., Kersten, S.R., van Swaaij, W.P., et al., 2009. Fast pyrolysis of biomass in a fluidized bed reactor: in situ filtering of the vapors. Industrial & Engineering Chemistry Research 48 (10), 4744–4756.

Hsu, D.D., 2012. Life cycle assessment of gasoline and diesel produced via fast pyrolysis and hydroprocessing. Biomass and Bioenergy 45, 41–47.

Huber, G.W., Corma, A., 2007. Synergies between bio-and oil refineries for the production of fuels from biomass. Angewandte Chemie International Edition 46 (38), 7184–7201.

Iribarren, D., Peters, J.F., Dufour, J., 2012. Life cycle assessment of transportation fuels from biomass pyrolysis. Fuel 97, 812–821.

Isahak, W.N.R.W., Hisham, M.W., Yarmo, M.A., Hin, T.Y.Y., 2012. A review on bio-oil production from biomass by using pyrolysis method. Renewable and Sustainable Energy Reviews 16 (8), 5910–5923.

Jahirul, M.I., Rasul, M.G., Chowdhury, A.A., Ashwath, N., 2012. Biofuels production through biomass pyrolysis—a technological review. Energies 5 (12), 4952–5001.

Kauffman, N., Hayes, D., Brown, R., 2011. A life cycle assessment of advanced biofuel production from a hectare of corn. Fuel 90 (11), 3306–3314.

Kimming, M., Sundberg, C., Nordberg, Å., Baky, A., Bernesson, S., Norén, O., et al., 2011. Biomass from agriculture in small-scale combined heat and power plants—a comparative life cycle assessment. Biomass and Bioenergy 35 (4), 1572–1581.
Koroneos, C., Dompros, A., Roumbas, G., 2008. Hydrogen production via biomass gasification—a life cycle assessment approach. Chemical Engineering and Processing: Process Intensification 47 (8), 1261–1268.
Lokesh, K., Ladu, L., Summerton, L., 2018. Bridging the gaps for a 'circular'bioeconomy: selection criteria, bio-based value chain and stakeholder mapping. Sustainability 10 (6), 1695.
Lu, H.R., El Hanandeh, A., 2019. Life cycle perspective of bio-oil and biochar production from hardwood biomass; what is the optimum mix and what to do with it? Journal of Cleaner Production 212, 173–189.
Luterbacher, J.S., Fröling, M., Vogel, F., Maréchal, F., Tester, J.W., 2009. Hydrothermal gasification of waste biomass: process design and life cycle asessment. Environmental Science & Technology 43 (5), 1578–1583.
Manfredi, S., Allacker, K., Pelletier, N., Chomkhamsri, K., de Souza, D.M., 2012. Product Environmental Footprint (PEF) Guide. European Commission - Joint Research Centre, Ispra, Italy.
McDowall, W., Anandarajah, G., Dodds, P.E., Tomei, J., 2012. Implications of sustainability constraints on UK bioenergy development: assessing optimistic and precautionary approaches with UK MARKAL. Energy Policy 47, 424–436.
Mirkouei, A., Haapala, K.R., Sessions, J., Murthy, G.S., 2017. A mixed biomass-based energy supply chain for enhancing economic and environmental sustainability benefits: a multi-criteria decision making framework. Applied Energy 206, 1088–1101.
Moghaddam, E.A., Ericsson, N., Hansson, P.A., Nordberg, Å., 2019. Exploring the potential for biomethane production by willow pyrolysis using life cycle assessment methodology. Energy, Sustainability and Society 9 (1), 1–18.
Mohan, D., Pittman Jr, C.U., Steele, P.H., 2006. Pyrolysis of wood/biomass for bio-oil: a critical review. Energy & Fuels 20 (3), 848–889.
Molino, A., Larocca, V., Chianese, S., Musmarra, D., 2018. Biofuels production by biomass gasification: a review. Energies 11 (4), 811.
Muench, S., Guenther, E., 2013. A systematic review of bioenergy life cycle assessments. Applied Energy 112, 257–273.
Neves, D., Thunman, H., Matos, A., Tarelho, L., Gómez-Barea, A., 2011. Characterization and prediction of biomass pyrolysis products. Progress in Energy and Combustion Science 37 (5), 611–630.
Nguyen, T.L.T., Hermansen, J.E., Nielsen, R.G., 2013. Environmental assessment of gasification technology for biomass conversion to energy in comparison with other alternatives: the case of wheat straw. Journal of Cleaner Production 53, 138–148.
Ningbo, G., Baoling, L., Aimin, L., Juanjuan, L., 2015. Continuous pyrolysis of pine sawdust at different pyrolysis temperatures and solid residence times. Journal of Analytical and Applied Pyrolysis 114, 155–162.
Oasmaa, A., Solantausta, Y., Arpiainen, V., Kuoppala, E., Sipila, K., 2010. Fast pyrolysis bio-oils from wood and agricultural residues. Energy & Fuels 24 (2), 1380–1388.
Oreggioni, G.D., Singh, B., Cherubini, F., Guest, G., Lausselet, C., Luberti, M., et al., 2017. Environmental assessment of biomass gasification combined heat and power plants with absorptive and adsorptive carbon capture units in Norway. International Journal of Greenhouse Gas Control 57, 162–172.
Pa, A., Bi, X.T., Sokhansanj, S., 2011. A life cycle evaluation of wood pellet gasification for district heating in British Columbia. Bioresource Technology 102 (10), 6167–6177.

Palomba, V., Frazzica, A., 2019. Comparative analysis of thermal energy storage technologies through the definition of suitable key performance indicators. Energy and Buildings 185, 88–102.

Pandey, B., Prajapati, Y.K., Sheth, P.N., 2019. Recent progress in thermochemical techniques to produce hydrogen gas from biomass: a state of the art review. International Journal of Hydrogen Energy 44 (47), 25384–25415.

Panoutsou, C., Singh, A., Christensen, T., Pelkmans, L., 2020. Competitive priorities to address optimisation in biomass value chains: the case of biomass CHP. Global Transitions 2, 60–75.

Peters, J.F., Iribarren, D., Dufour, J., 2015a. Biomass pyrolysis for biochar or energy applications? A life cycle assessment. Environmental Science & Technology 49 (8), 5195–5202.

Peters, J.F., Iribarren, D., Dufour, J., 2015b. Simulation and life cycle assessment of biofuel production via fast pyrolysis and hydroupgrading. Fuel 139, 441–456.

Pütün, A.E., Önal, E., Uzun, B.B., Özbay, N., 2007a. Comparison between the "slow" and "fast" pyrolysis of tobacco residue. Industrial Crops and Products 26 (3), 307–314.

Pütün, A.E., Özbay, N., Apaydın Varol, E., Uzun, B.B., Ateş, F., 2007b. Rapid and slow pyrolysis of pistachio shell: effect of pyrolysis conditions on the product yields and characterization of the liquid product. International Journal of Energy Research 31 (5), 506–514.

Puy, N., Rieradevall, J., Bartrolí, J., 2010. Environmental assessment of post-consumer wood and forest residues gasification: the case study of Barcelona metropolitan area. Biomass and Bioenergy 34 (10), 1457–1465.

Ren, J., Liu, Y.L., Zhao, X.Y., Cao, J.P., 2020. Biomass thermochemical conversion: a review on tar elimination from biomass catalytic gasification. Journal of the Energy Institute 93 (3), 1083–1098.

Rezaei, H., Lim, C.J., Lau, A., Sokhansanj, S., 2016. Size, shape and flow characterization of ground wood chip and ground wood pellet particles. Powder Technology 301, 737–746.

Righi, S., Bandini, V., Marazza, D., Baioli, F., Torri, C., Contin, A., 2016. Life cycle assessment of high ligno-cellulosic biomass pyrolysis coupled with anaerobic digestion. Bioresource Technology 212, 245–253.

Roberts, K.G., Gloy, B.A., Joseph, S., Scott, N.R., Lehmann, J., 2010. Life cycle assessment of biochar systems: estimating the energetic, economic, and climate change potential. Environmental Science & Technology 44 (2), 827–833.

Salkuyeh, Y.K., Saville, B.A., MacLean, H.L., 2018. Techno-economic analysis and life cycle assessment of hydrogen production from different biomass gasification processes. International Journal of Hydrogen Energy 43 (20), 9514–9528.

Sansaniwal, S.K., Pal, K., Rosen, M.A., Tyagi, S.K., 2017. Recent advances in the development of biomass gasification technology: a comprehensive review. Renewable and Sustainable Energy Reviews 72, 363–384.

Sarangi, P.K., Nanda, S., Mohanty, P. (Eds.), 2018. Recent advancements in biofuels and bioenergy utilization, Vol. 232. Springer, Berlin, Germany.

Shahabuddin, M., Krishna, B.B., Bhaskar, T., Perkins, G., 2020. Advances in the thermochemical production of hydrogen from biomass and residual wastes: summary of recent techno-economic analyses. Bioresource Technology 299, 122557.

Shemfe, M.B., Whittaker, C., Gu, S., Fidalgo, B., 2016. Comparative evaluation of GHG emissions from the use of Miscanthus for bio-hydrocarbon production via fast pyrolysis and bio-oil upgrading. Applied Energy 176, 22–33.

Steubing, B., Zah, R., Ludwig, C., 2011. Life cycle assessment of SNG from wood for heating, electricity, and transportation. Biomass and Bioenergy 35 (7), 2950–2960.

Susmozas, A., Iribarren, D., Zapp, P., Linßen, J., Dufour, J., 2016. Life-cycle performance of hydrogen production via indirect biomass gasification with CO_2 capture. International Journal of Hydrogen Energy 41 (42), 19484–19491.
Talmadge, M.S., Baldwin, R.M., Biddy, M.J., McCormick, R.L., Beckham, G.T., Ferguson, G.A., et al., 2014. A perspective on oxygenated species in the refinery integration of pyrolysis oil. Green Chemistry 16 (2), 407–453.
Thornley, P., Gilbert, P., Shackley, S., Hammond, J., 2015. Maximizing the greenhouse gas reductions from biomass: the role of life cycle assessment. Biomass and Bioenergy 81, 35–43.
Tripathi, M., Sahu, J.N., Ganesan, P., 2016. Effect of process parameters on production of biochar from biomass waste through pyrolysis: a review. Renewable and Sustainable Energy Reviews 55, 467–481.
Tsai, W.T., Lee, M.K., Chang, D.Y., 2006. Fast pyrolysis of rice straw, sugarcane bagasse and coconut shell in an induction-heating reactor. Journal of Analytical and Applied Pyrolysis 76 (1–2), 230–237.
Uzun, B.B., Pütün, A.E., Pütün, E., 2006. Fast pyrolysis of soybean cake: product yields and compositions. Bioresource Technology 97 (4), 569–576.
van Dyk, S., Su, J., Mcmillan, J.D., Saddler, J., 2019. Potential synergies of drop-in biofuel production with further co-processing at oil refineries. Biofuels, Bioproducts and Biorefining 13 (3), 760–775.
Venderbosch, R.H., Ardiyanti, A.R., Wildschut, J., Oasmaa, A., Heeres, H.J., 2010. Stabilization of biomass-derived pyrolysis oils. Journal of Chemical Technology & Biotechnology 85 (5), 674–686.
Vienescu, D.N., Wang, J., Le Gresley, A., Nixon, J.D., 2018. A life cycle assessment of options for producing synthetic fuel via pyrolysis. Bioresource Technology 249, 626–634.
Wang, S., Dai, G., Yang, H., Luo, Z., 2017. Lignocellulosic biomass pyrolysis mechanism: a state-of-the-art review. Progress in Energy and Combustion Science 62, 33–86.
Welcomer, S., 2008. The Sustainability Handbook: The Complete Management Guide to Achieving Social, Economic and Environmental Responsibility. JSTOR.
Wiloso, E.I., Heijungs, R., De Snoo, G.R., 2012. LCA of second generation bioethanol: a review and some issues to be resolved for good LCA practice. Renewable and Sustainable Energy Reviews 16 (7), 5295–5308.
Winjobi, O., Shonnard, D.R., Bar-Ziv, E., Zhou, W., 2016. Life cycle greenhouse gas emissions of bio-oil from two-step torrefaction and fast pyrolysis of pine. Biofuels, Bioproducts and Biorefining 10 (5), 576–588.
Wong, A., Zhang, H., Kumar, A., 2016. Life cycle assessment of renewable diesel production from lignocellulosic biomass. International Journal of Life Cycle Assessment 21 (10), 1404–1424.
Yu, H., Román, E., Solvang, W.D., 2018. A value chain analysis for bioenergy production from biomass and biodegradable waste: a case study in northern Norway. Energy Systems and Environment 183.
Zhang, Q., Chang, J., Wang, T., Xu, Y., 2007. Review of biomass pyrolysis oil properties and upgrading research. Energy Conversion and Management 48 (1), 87–92.

Further reading

Bach, V., Lehmann, A., Görmer, M., Finkbeiner, M., 2018. Product environmental footprint (PEF) pilot phase—comparability over flexibility? Sustainability 10 (8), 2898.

Cabeza, L.F., Galindo, E., Prieto, C., Barreneche, C., Fernández, A.I., 2015. Key performance indicators in thermal energy storage: survey and assessment. Renewable Energy 83, 820–827.

Christoforou, E., Fokaides, P.A., 2018. Advances in Solid Biofuels. Springer.

Doublet, G., Ingólfsdóttir, G.M., Yngvadóttir, E., Landquist, B., Jungbluth, N., Aronsson, A., et al., 2014. Key Environmental Performance Indicators for a simplified LCA in food supply chains. In: 9th International Conference LCA of Food, San Francisco, pp. 8–10.

Finnveden, G., Hauschild, M.Z., Ekvall, T., Guinée, J., Heijungs, R., Hellweg, S., et al., 2009. Recent developments in life cycle assessment. Journal of Environmental Management 91 (1), 1–21.

Fokaides, P.A., Christoforou, E., 2016. Life cycle sustainability assessment of biofuels. Handbook of Biofuels Production. Woodhead Publishing, pp. 41–60.

Jasch, C., 2000. Environmental performance evaluation and indicators. Journal of Cleaner Production 8 (1), 79–88.

Purvis, B., Mao, Y., Robinson, D., 2019. Three pillars of sustainability: in search of conceptual origins. Sustainability Science 14 (3), 681–695.

Wolf, C., Klein, D., Weber-Blaschke, G., Richter, K., 2016. Systematic review and meta-analysis of life cycle assessments for wood energy services. Journal of Industrial Ecology 20 (4), 743–763.

Zampori, L., Pant, R., 2019. Suggestions for updating the Product Environmental Footprint (PEF) method. JRC Technical Reports; Publications Office of the European Union: Luxembourg, p. 76.

CHAPTER 6

Environmental assessment of biomass to biofuels: biochemical conversion routes

Guillermo Garcia-Garcia[1,2], Stephen McCord[1] and Peter Styring[1]
[1]Department of Chemical and Biological Engineering, The University of Sheffield, Sheffield, United Kingdom
[2]Department of Agrifood System Economics, Centre 'Camino de Purchil', Institute of Agricultural and Fisheries Research and Training (IFAPA), Granada, Spain

Contents

6.1 Introduction

Biomass is any material composed of plant or animal matter, such as manure, wood, household waste, and crops. Biomass can be used to generate heat and/or electricity or as a fuel for transportation, and as such, it can be considered a renewable energy source. Because of this, there is a growing interest in substituting conventional fossil-based energy sources, such as coal and petrol, with renewable cleaner alternatives, such as

Environmental Assessment of Renewable Energy Conversion Technologies
DOI: https://doi.org/10.1016/B978-0-12-817111-0.00010-3

biomass. The energy obtained by biomass can be considered a form of bioenergy (Chum et al., 2011).

Biomass can be used directly to produce energy, for example, by burning wood logs. However, biomass is often treated to obtain a product that is more suited for energy generation. Some of these treatments include thermochemical processes, such as pyrolysis and gasification, which use very high temperatures. An alternative to this is biochemical processes, in which biomass transforms into biofuels via biochemical reactions with the help of microorganisms (Fig. 6.1).

Biofuels are classified into four major generations (Fokaides and Christoforou, 2016). First-generation biofuels are obtained from food crops grown on arable lands such as wheat, corn, sugar, and beet. Second-generation biofuels are made from nonfood crops, agricultural waste, forest residues, and industrial wastes. Third-generation biofuels are produced from algae. Finally, fourth-generation biofuel uses genetically modified algae or cyanobacteria to optimize biofuel production. Commercialization is dominated by second-generation biofuels, while the third- and fourth-generation are generally not economically viable yet at a large scale (Abdullah et al., 2019). The main biofuels that can be produced via any of the four generations are biogas, bioethanol, and biodiesel. Biogas is mainly used to generate heat and/or electricity, while bioethanol and biodiesel are mostly used in vehicle engines for transportation. Other biofuels include other bioalcohols (e.g., biomethanol, biopropanol, biobutanol, and bioglycerol) and bioethers (e.g., dimethyl ether, diethyl ether, and methyl *tert*-butyl ether). This chapter focuses on biogas and bioethanol produced via biochemical conversion routes, as opposed to biodiesel, generally produced via chemical reactions of transesterification or other chemicals produced via thermochemical routes.

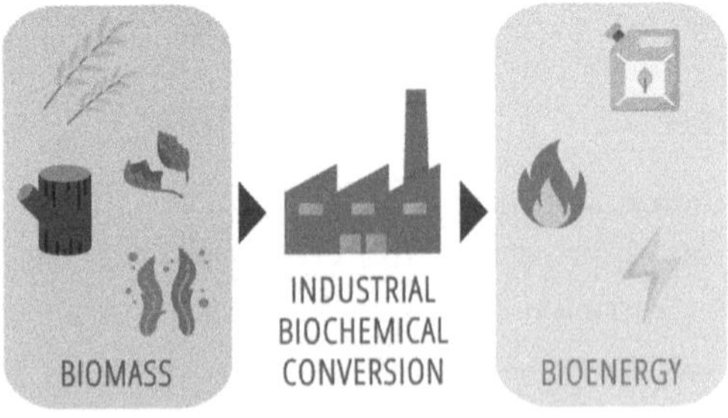

Figure 6.1 Biochemical conversion of biomass to biofuels to be used to obtain bioenergy.

Demand for such products is rapidly increasing. The World Biogas Association reports that there are around 132.000 small, medium, or large-scale digesters and 50 million micro-scale digesters operating in the world (Jain et al., 2019). The IEA stated that global biogas production was around 35 Mtoe in 2018 (International Energy Agency, 2020). This is only a small fraction of the potential production, all sustainable feedstocks should be digested—a figure the IEA reports as 570 Mtoe as of 2018, equal to about 20% of global natural gas demand. A range of feedstocks is viable, including crop biomass, municipal organic waste, wastewater, and animal manure. The vast majority of the biogas produced is used to generate electricity and/or heat. Global electricity generation from biogas nearly doubled from 2010 to 2016 (Jain et al., 2019).

Liquid biofuel production reached 163 billion liters/year in 2019, with expected production between 2023 and 2025 estimated to average 182 billion liters/year (International Energy Agency, 2020). Bioethanol is the most used biofuel worldwide, with its production reaching 110−115 billion liters globally in 2019, with an expected 119 billion liters in 2023−25 (Susmozas et al., 2020; International Energy Agency, 2020). The United States and Brazil are the clear leaders in the bioethanol market. Furthermore, the production of biogasoline (such as bioethanol) increased by 1.8% in 2019, mostly due to a production increase in the Asia Pacific, Canada, Mexico, and Brazil (BP, 2020). Its consumption increased by 4.1% worldwide and by 9.2% in the European Union in 2019.

Nevertheless, the production of biofuels from biomass still faces several challenges. In spite of the numbers mentioned above, modern biofuels currently represent a very minor proportion of all the energy production in the world. Until the end of the 19th century, most of the energy was produced from biomass burning such as wood, crop waste, or charcoal. However, nowadays, most energy worldwide is generated from coal, oil, and gas. Some challenges to overcome are related to the technology and cost of biofuel production and their competition with conventional fossil fuels. First-generation biofuels present the additional issue of diverting crops that could be used for human consumption to biofuels production. This "food versus fuel" dilemma may have an important effect on food prices.

In addition to the considerations made above, the environmental impact of biofuel production must also be taken into account. Although biofuel is a form of renewable energy, its production and use have associated an environmental impact. That is why methodologies like life cycle assessment (LCA) must be used to analyze the environmental impact of biofuels over their entire life cycle.

This chapter starts by providing an overview of the state-of-the-art technologies used to convert biomass into biofuels, introduces the LCA methodology and its use to analyze biofuels, then presents environmental impact results for biofuels, and finally discusses the Product Environmental Footprint (PEF) method for measuring sustainability performance of biofuels.

6.2 State-of-the-art of the production technologies

There are two dominant biochemical conversion routes used for biofuel production: fermentation and anaerobic digestion (AD), which are reviewed below.

6.2.1 Fermentation

Liquid biofuels play an important role within the supply of global transport fuels. The most widely produced liquid biofuel is bioethanol, most of which is produced through fermentation processes. Two countries dominate global bioethanol production: the United States and Brazil. In the United States, 9.5 billion liters of bioethanol were produced in 2019, 94% of which were derived from pure corn feedstock and a further 5.9% from corn mixed with a secondary material (sorghum or cellulosic biomass) (RFA, 2020). In Brazil, 36 billion liters of bioethanol were produced in 2019, 99.2% of which was derived from sugarcane, with the remaining fraction made up of corn and bagasse (Barros, 2019).

Therefore, current bioethanol production is dominated by first-generation processes, although an increased focus has been placed on developing second- and third-generation alternatives. Regardless of the feedstock, most biomass fermentation processes contain the same process stages: pretreatment, hydrolysis/saccharification, fermentation, and distillation/separation.

Pretreatment prepares the biomass source for hydrolysis and fermentation. Simply, pretreatment strategies aim to extract as much fermentable material as possible from the feedstock to increase the overall yield and profitability. A number of objectives for this stage can be ascertained (Rastogi and Shrivastava, 2017; Balan, 2014)

- Reduce the crystallinity and polymerization of cellulose, as well as increase the porosity of the lignocelluloses to enhance sugar yields during hydrolysis.
- Limit the degradation of sugars derived from cellulose and hemicellulose.
- Minimize the formation of products that are known to inhibit fermentation.
- Recover lignin for conversion into co-products or disposal.
- Be cost-effective by minimizing energy inputs whilst maximizing sugar yield and try to use cheap and/or less hazardous chemicals where possible.
- Recover the catalyst for environmental and economic reasons.

Most pretreatment methods are physical or chemical in nature, although biological processes can also be used, and in some instances, a mixture of pretreatment methods may be used. In the market dominated by corn and sugarcane plants, mechanical processes are typical. Ninety percent of bioethanol in the United States is produced through dry milling, where the pretreatment consists of grinding the whole grain before it is slurried with water. For sugarcane in Brazil, pretreatment consists of crushing the cane stalk in an extractor/expeller to extract the sugar-rich cane juice.

However, other treatment forms are considered for many second-generation feedstocks, with the functional effectiveness depending on both the treatment itself and the biomass treated. An example of this is provided by Balan (2014), who stated that ammonia fiber expansion is effective on monocots such as sorghum, sugarcane, and bamboo but not on dicots such as poplar and willow. Compositional changes in the plant cell wall and associated ultrastructure are known to influence the effectiveness of pretreatment, with this affected by species and environmental conditions during the growing season, further complicating the issue.

Table 6.1 provides a summary of selected pretreatment options.

The hydrolysis/saccharification process stage follows pretreatment. This stage can be handled separately to fermentation [separate hydrolysis and fermentation (SHF)] or simultaneously [saccharification and fermentation/co-fermentation (SSF/SSCF)]. SHF processes offer the advantage of being able to operate both processes at their optimum conditions. However, this is balanced against a more complex plant and the accumulation of sugars inhibiting enzyme activity during hydrolysis. For SSF/

Table 6.1 Comparison of some pretreatment methods in biomass fermentation (Houfani et al., 2020; Rastogi and Shrivastava, 2017).

Pretreatment	Description	Advantages	Disadvantages
Mechanical	Utilizes processes such as milling, grinding, shredding, and chipping to reduce biomass particle size and disrupt surface structure using either compressive or shear forces	■ Increases surface area to increase digestibility	■ High energy-consumption and low efficiency result in high costs
Liquid hot water	Pressurized (5 MPa) hot water (160°C–240°C) is used to degrade hemicellulose	■ Increased cellulose availability ■ No chemicals/solvents	■ High energy and water consumption
Steam explosion	Biomass is exposed to hot steam (160°C–240°C) under high pressure. The pressure is released rapidly, resulting in auto-hydrolysis of hemicellulose, which can be run as a catalytic or noncatalytic process	■ High sugar yield ■ Less hazardous chemical requirements ■ Improvements to enzymatic hydrolysis	■ Effectiveness is dependent on biomass source ■ Can form chemicals that inhibit other fermentation stages ■ Can degrade hemicellulose and lignin ■ High costs (energy)
Ammonia fiber expansion	Biomass is treated with NH_3 with an elevated temperature (60°C–100°C) and pressure (17.0–20.7 MPa). The pressure is released rapidly, resulting in disruption of fibers and partial de-crystallization of cellulose	■ Increases surface area to increase digestibility ■ Low rate of formation for toxic chemicals or those that inhibit fermentation	■ Not as effective on high lignin biomass sources ■ High costs (NH_3) ■ Lower sugar yield-hemicellulose is not reduced as effectively

Acid	Dilute acid is used to increase the solubility of hemicellulose and lignin. The acid used can either be at a higher temperature (~180°C) for a short retention time or at a lower temperature (~120°C) for a longer one	■ Enhanced hydrolysis of hemicellulose and amorphous cellulose ■ High sugar yield	■ High costs (energy) ■ Presence of corrosive acids ■ Can form chemicals that inhibit other fermentation stages
Alkali	Hydroxides such as NaOH, KOH, or NH_4OH are used to disrupt ester and glycosidic side-chains resulting in the alteration of lignin structure and leading to partial de-crystallization of hemicellulose	■ Efficient lignin removal ■ Increased hemicellulase accessibility ■ Decreased biomass crystallinity	■ Downstream process costs are higher than alternative routes
Organosolv	Organic solvents are mixed with inorganic acid catalysts to boost lignin extraction	■ Improved digestibility of lignin and hemicellulose	■ High costs (solvents, catalysts) ■ Hazardous (risk of fires/explosions)
Biological	Microorganisms such as white-rot fungi are used that produce enzymes capable of degrading (e.g., lignin)	■ Low-capital-cost ■ Low operational cost: low energy, no chemicals required, mild reaction conditions	■ Slow rate of hydrolysis when compared to alternatives

SSCF the reverse is true: sugars are processed quickly and the resultant ethanol helps to prevent contamination. Nonetheless, there is a difficulty in optimizing both process steps simultaneously due to variance in preferred operating conditions for the enzymes and the microbes for fermentation. Existing corn and sugarcane processes are typically SSF processes with one reactor for both process stages.

Typically, enzymes are delivered within a "free enzyme" system as this is easier to duplicate than the "cellulosomal" alternative. The enzymes are produced commercially using fungi or bacteria and consist of a mixture of cellulases, hemicellulases, and pectinases. Each enzyme acts on different active sites and produces different products; see Table 6.2 for examples. Since enzymes are expensive, there is a need to recover and recycle them, and a variety of processes are available (Houfani et al., 2020).

After hydrolysis/saccharification, fermentation takes place, converting sugars into a range of products, most frequently ethanol, although other alcohols or chemicals such as organic acids can be produced. The chemical produced, and the yield is dependent on the microbe strain used. A number of strains are shown in Table 6.3. While bioethanol processes dominate the commercial space, currently, acetone-butanol-ethanol (ABE) processes have a long history of prior commercial success, and new developments have seen the opening of new plants (ABE, biobutanol) in many locations across the globe (Veza et al., 2021).

After fermentation, the product broth requires separation. Enzymes and microbial matter are recovered through filtration for sterilization and reuse, while distillation is the preferred commercial method for product purification. For ethanol/water mixes, molecular sieves or additives are needed to pass through any azeotrope regions, for example, binary ethanol-water or tertiary mixtures of ethanol-water-butanol in acetone-butanol-ethanol (ABE) fermentation. It should be noted that for ABE processes, in particular, alternative separation processes may be more valid.

6.2.2 Anaerobic digestion

AD is the second major biochemical conversion pathway in the production of biofuels. AD processes are capable of turning a range of biomass sources into biogas, a mixture of predominantly CH_4 and CO_2.

AD consists of several stages: hydrolysis, acidogenesis, acetogenesis, and methanogenesis. These stages take place typically within a singular reactor, although two-stage AD systems can be operated. Pretreatment and biogas

Table 6.2 Selected enzymes used in the treatment of lignocellulosic biomass (Houfani et al., 2020).

Enzyme	EC number	Active site	Hydrolysis products
Cellulases			
Endoglucanase	EC 3.2.1.4	Randomly on the cellulose chain	Cello-oligomers
Cellobio hydrolase	EC 3.2.1.91	Reducing end of cellulose	Cellobiose
Cellobiohy drolase	EC 3.2.1.176	Non-reducing end of cellulose	Cellobiose
β-glucosidase	EC 3.2.1.21	Celloboise	Glucose
LPMO		C1 or C4 of glucose	Aldonic acid
Depolymerization Hemicellulases			
Endo-1, 4-β-xylanase	EC 3.2.1.8	β-1,4 xylan Exo	Xylo-oligomers
β-xylosidase	EC 3.2.1.37	Exo β-1,4 xylo-oligomers	Xylose
Endo-1, 4-mannanase	EC 3.2.1.78	β-1,4 mannane	Mannans oligomers
β-mannosidase	EC 3.2.1.25	β-1,4 mannans oligomers	Mannose
Accessory Hemicellulases			
α-L-arabinof uranosidase	EC 3.2.1.55	Positions O-2 and O-3 des α-L-arabin ofuranosyles	L-arabinose
α-L-arabinanase	EC 3.2.1.99	α-1,5-arabinan	Arabinose
α-alactosidases	EC 3.2.1.22	Oligosaccharides with α galactosyl groups	Galactose and sucrose

upgrading are optional process stages. AD processes can be wet or dry, with the choice of system dependent on the dry matter content of the feed.

Table 6.3 Examples of microbial strains used for the fermentation of biomass for biofuel production (Balan, 2014).

Strain	Description	Product	Yield (g/g consumed sugar)
Saccharomyces cerevisiae TMB3400	Modified strain for fermenting xylose and glucose	Ethanol	0.4
S. cerevisiae RWB 218	Modified strain for fermenting xylose and glucose	Ethanol	0.38
Escherichia coli KO11	Homoethanolic fermentation strain (modified)	Ethanol	0.34–0.4
Zymomonas mobilis AX101	Genetically engineered xylose, arabinose, and glucose-fermenting strain	Ethanol	0.42
Clostridium thermocellum LQRI	Native consolidated bioprocessing strain	Ethanol	0.26
Clostridium acetobutylicum P262	Native acetone, butanol, and ethanol-producing strain	ABE	0.39
Rhodosporidium toruloides Y4	Lignocellulosic hydrolysate domesticated strain could ferment glucose and xylose	Lipids	0.1

Pretreatment is optional but can be used to improve substrate digestibility and availability in the AD process. As with fermentation, pretreatment can be physical, chemical, biological, or combined in nature. A number of alternatives under development are reported in Table 6.4.

In the digestion stage, the constituent proteins, carbohydrates, and lipids of the feedstock are converted into biogas through a range of intermediates in a variety of complex pathways driven by a mixture of bacteria. Hydrolysis serves the purpose of transforming larger macromolecules into smaller components that can be utilized by acidogenic bacteria, with enzymes excreted by hydrolytic bacteria responsible for this. Acidogenesis sees acidogenic bacteria absorb hydrolysis products and secrete volatile fatty acids such as acetates but also larger organic acids like propionate and butyrate. The larger acids are converted to acetate through acetogenesis, with lipids also converted through their own acetogenic process. Hydrogen is also produced in this stage. Methanogenesis sees the generation of methane by methanogenic species through either the acetoclastic route (66.6% of total methane production utilizing acetate as a feed) or

Table 6.4 Selected AD pretreatment options in development and their reported impacts on biogas production (Atelge et al., 2020).

Pretreatment	Substrate	Process	Impact on AD process
Mechanical: Hollander beater	Laminariaceae spp	Reduce particle size to 76–836 microns	Methane yield up 53%
Mechanical: shockwave	Hemp fibers	Pressure shockwaves of 60 MPa applied continuously	20% increase in COD removal efficiency
Mechanical: metal plates	Ensiled grass	Application of two coarse meshes for grating material	Methane yield up 12.3%
Thermal pretreatment	Sludge, MSW, spent grain, and cow manure	Heating of feed material to 150°C–170°C for 30 min	Methane yield up by 30%, 70% faster MSW degradation kinetics
Thermal: steam explosion	Late harvest hay	Treatment with steam at 160°C–220°C for 5–15 min	Methane yield up 16%
Thermal hydrolysis	Sewage sludge	Hydrolysis treatment at 160°C for 30 min under 6 bar of pressure	Improved volatile solid removal and availability for AD
Ultrasonic pre-treamtent	Food waste	Treatment with a specific amount of ultrasonic energy for a specified time period (in this case 130 W at 20 kHz, 30 min)	Biogas yield up 10.12%, improved removal of solids, volatile solids, and COD
Alkaline pretreatment	Rice straw	Treatment with hydroxide (in this case sodium) at a moderate temperature (30°C) for 24 h	Methane yield up 21.4%
Ozonation	Sugarcane bagasse	Treatment with ozone (0.4–0.875 mgO_3/g TS) for 15 min	Improved lignin (45.2%) and hemicellulose (48.28%) removal
Fenton pretreatment	Olive mill waste	Fenton process pretreatment of wastewater from an olive mill	88.8% of total phenolic content removed, 35.4% of TOC removed

the hydrogenotrophic route (33.3% of total, hydrogen-fed). A number of techniques are available for biogas upgrading, primarily intending to produce a purer stream of "biomethane" with organic membranes being a common choice.

Current research focuses primarily on improving biogas yield and quality. Accelerating the two stages often determined to be rate determining (hydrolysis and methanogenesis) is one area of focus. Techniques such as microaeration (dosing small amounts of air) have been shown to accelerate hydrolysis and scavenge H_2S (Chen et al., 2020). Other studies have looked at the impact of changing biomass composition, as detailed by Rasapoor et al. (2020), mixing protein-rich biomass (e.g., manure) with carbon sources (e.g., crop wastes) to increase the biogas yields. The same source also provides data on the impact of adding materials such as activated carbon or biochar into the digester resulting in greater CH_4 production. In both cases, many reported results show an increase in excess of 100%.

6.3 Calculation of environmental impacts via life cycle assessment

A biofuel, like any other product, has associated an environmental impact. Importantly, this environmental impact is not only generated during production. It is paramount to consider the impacts generated in all life cycle stages of a product: extraction of raw materials, production, distribution, use, and end-of-life. Such life cycle stages for biofuel production from biomass are represented in Fig. 6.2. LCA is a methodology to quantify the environmental impacts of a product, process, company, etc. over its entire life cycle. LCA accounts for the consumption of resources, emissions to air, water, and soil, as well as the treatment of any waste generated. Once all these data are collected, LCA allows calculating the environmental impacts associated with those material flows.

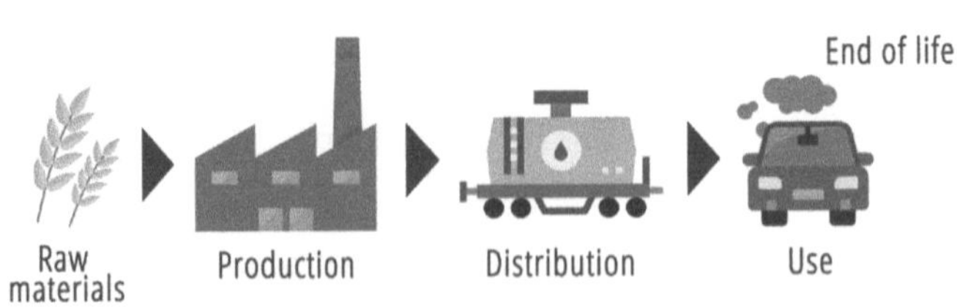

Figure 6.2 Life cycle stages of a generic biofuel.

Due to its increasing use, LCA has been standardized by the International Organization for Standardization (ISO). ISO 14040:2006 (ISO 14040, 2006) describes the principles and framework for LCA, while 14044:2006 (ISO 14044, 2006) specifies requirements and provides guidelines for the phases and areas described in ISO 14040:2006 (ISO 14040, 2006). ISO divides an LCA into four major phases: definition of the goal and scope, life cycle inventory analysis, life cycle impact analysis, and interpretation. Each of these four phases and their application to the production of biofuels from biomass is explained below.

6.3.1 Definition of the goal and scope

In the first step of any LCA, the goal of the study must be defined. The goal should disclose the intention and reason for the study, the target audience, the commissioners, and the authors. It must be clear, unambiguous, and specific to a location and time span. During this step, the functional unit (FU) is defined. FU is used as a reference, on which all the mass and energy flows are based, allowing an accurate comparison between technologies. It should consider the performance of the product studied. An example of a FU for biofuel is "to compare the life cycle environmental impact of a car model X driving for 100 km fueled by E100 bioethanol produced from maize grown in Brazil with the same car fueled by the conventional gasoline used in Brazil in 2021." The reference flow is the mass flow of the final product required to fulfill the function defined by the FU. In a hypothetical example, in which E100 bioethanol was 10% more efficient than conventional gasoline, the reference flows would be the production of "0.9 *Y* liters E100 bioethanol" and "*Y* liters gasoline" respectively, allowing for a fair comparison of both products.

A key aspect of the study's scope is defining the boundaries of the system to be analyzed. This means that the inclusion and exclusion of processes must be decided. At this point, some stages of the life cycle may be excluded if they are not relevant for the study goal or data are not available. For instance, cradle-to-grave, cradle-to-gate, or gate-to-gate approaches may be used.

Care should also be taken to consider the impact allocation for all generations of biofuels. In many instances, biochemical routes can utilize the waste materials (e.g., second-generation corn stover bioethanol) or produce co-products with significant value (e.g., first-generation corn bioethanol, ABE process), leaving the practitioner a choice on how to assign burdens to each product. For these multifunctional processes, it is recommended to apply subdivision or system expansion. Should allocation be necessary, then three options exist

1. Main production: where all impacts are allocated to the main product.
2. Physical: where impacts are allocated based on physical properties (mass, volume).
3. Economic: where impacts are allocated based on the economic value of products.

Other aspects to define in this stage include the type of impact assessment methodology and impact categories used in later phases of the LCA. Table 6.5 lists the relevant and recent LCA studies of biomass to biofuels via biochemical routes, showing the scope of the study.

6.3.2 Life cycle inventory analysis

This is the stage in which most of the data are collected. All emissions to air, water, and soil must be identified and quantified for all processes identified within the system boundaries. Additionally, any waste material that is sent to waste treatment or recycling process must be defined, along with other by-products produced. These are the outputs of the processes. The resources used by these processes, which are the inputs, must also be listed. These resources may come from nature or the technosphere. The quality of the data collected must also be disclosed, including factors such as technological and geographical representativeness, completeness, and reliability.

This is usually the stage that requires more time to complete due to the large amounts of data needed. The data can be collected via empirical measurements, simulation, estimation using expert knowledge, review of the literature, and commercial databases. Table 6.6 shows a list of data that must be collected for an exemplified production of bioethanol from forest residues. It must be noted that specific values are needed for all categories on the right column, in the following or equivalent units: kilowatt-hour (kWh) for electricity, megajoules (MJ) for heat, cubic meters (m^3) for water, and kilograms (kg) for the rest of the categories.

6.3.3 Life cycle impact assessment

Once the life cycle inventory data have been collected, these must be used to calculate the environmental impact of the different processes. Due

Table 6.5 Recent LCA studies of biomass to biofuels via biochemical routes.

Scope of the study	References
Anaerobic mono- and codigestion of residue grass to produce biogas	Bedoić et al. (2019)
Codigestion of pig manure and ensiled meadow grass biomass for biogas production	Tsapekos et al. (2017)
Codigestion of straw, sugar beet, and manure to produce natural gas, heat, and electricity	Jensen et al. (2017)
AD of lignocellulosic biomass for biogas production	Abraham et al. (2020)
AD of wetland biomass for biogas production	Roj-Rojewski et al. (2019)
Production of biogas from microalgae	Zabed et al. (2020)
Ethanol production from sugarcane, corn, and sugar beet	Manochio et al. (2017)
Ethanol production from lignocellulosic biomass	Moreno et al. (2017)
Ethanol production from crop residues and weedy biomass	Pandiyan et al. (2019)
Ethanol production via cyanobacterial biomass and microalga	Rizza et al. (2019)
Ethanol production via engineered cyanobacteria	Liang et al. (2018)
Butanol production from biomass	Abo et al. (2019)
Smart fermentation engineering for butanol production from biomass	Zhao et al. (2019)

to the amount of data and the complexity of calculations, this is typically done with software such as SimaPro, GaBi, or openLCA.

There are different steps within this LCA phase. Mandatory stages, according to ISO standards, are classification and characterization. Different environmental impact categories would be considered based on the impact analysis methodology selected in the goal and scope definition. Important impact categories for biofuel production from biomass include global warming potential (GWP), land use, water use, human toxicity, and resource depletion. Each input and output identified in the inventory is classified according to the impact they produce during classification. For example, when biofuel is combusted and generates a CO_2 emission, this is classified in the climate change category. Each input and output is multiplied by a characterization factor during characterization, depending on how much they affect their impact category. This is done to calculate the value of each impact category in the same unit and able to aggregate and

Table 6.6 Example of the data needed to collect a life cycle inventory of bioethanol production from forest residues.

Reference flow			Bioethanol
Avoided products			Mixed alcohols, sulfur
Inputs	**From nature**		Oxygen, nitrogen (atmospheric), water
	From technosphere	**Materials**	Forest residue, diesel, gasoline, sulfite, silica sand, magnesium oxide, zeolite powder, monoethanolamine, hydrochloric acid, and sodium hydroxide
		Non-materials	Electricity, heat
Emissions	**To air**		Ammonia, carbon dioxide, nitrogen, oxygen, sulfur dioxide, nitrogen dioxide, and water
	To water		
	To soil		
Waste			Inert material to sanitary landfill, wood ash mixture to sanitary landfill, and sewage to wastewater treatment

Source: Data categories from United States life cycle inventory database (National Renewable Energy Laboratory, 2015).

sum results. For example, the GWP for 100 years of 1 kg of CO_2 emitted to the atmosphere is 1 kg CO_2eq, while for methane and nitrous oxide, they are 28 CO_2-eq and 298 CO_2-eq, respectively.

Once the steps above are completed, the environmental impacts of the process can be calculated in units such as kilograms of CO_2 (for GWP), m^2 (for land use), cubic meters (for water use), comparative toxic unit (for human toxicity), and kilograms Sb-eq (for resource depletion). Next, the following optional steps can be taken to more easily interpret results: normalization, grouping, weighting, and single score. Normalization allows identifying which environmental impact categories contribute the most to the total environmental impact by comparing each result to a reference value. For example, GWP may be more important than ozone depletion for biofuels. Grouping and weighting are used to group the impact categories together and assign them a weight proportional to their importance, according to the LCA practitioner. Finally, all values obtained can be aggregated into a single overall score, so different scenarios, processes, or products can be easily compared.

Since some of these steps demand a lot of calculations, while others are also relatively subjective (e.g., weighting), it is recommended to use an existing life cycle impact assessment methodology. The most up-to-date and recommended methodologies are ReCiPe, CML, EPD, and ILCD. Some of these methodologies follow a midpoint approach, in which results are reported in physical units (e.g., kg CO_2-eq). In contrast, others use an endpoint approach, in which results are based on the damage caused, using units such as disability-adjusted life years for impacts to humans, species.yr for impacts to ecosystems, and US dollars 2013 for resource scarcity.

6.3.4 Interpretation

In the interpretation phase, the results are reported and discussed. At this stage, it is important to refer back to the study's goal and scope to assess whether the objectives of the study have been met. Often, there is a need to revise some of the previous LCA phases and make adjustments. This makes LCA an iterative process. Finally, the conclusions and recommendations of the study must be provided.

The LCA report should also include an analysis and discussion of the model's quality of results and robustness. For this, several methods may be used, like completeness analysis, consistency check, sensitivity and uncertainty analysis, and identification of the most relevant issues.

6.4 Key performance indicators for life cycle assessment

A variety of feedstocks, locations, products, and technologies are all plausible for use in biochemical processes. While this flexibility is beneficial in many respects, one area in which it is somewhat problematic is in impact assessment, including LCA. In particular, this is problematic for comparative studies, where one may be interested in comparing the performance of one alternative against another in the same scenario.

One way to simplify this issue is to identify a number of key performance indicators (KPIs) that can provide a reasonable indication of environmental performance. In this sense, ideal KPIs can identify the elements of a process that contribute most significantly to the impact scores calculated in the LCA. The most basic KPIs are those related to the output product in the form of environmental impact measure per unit of product, for example, kg CO_2-eq/kg ethanol. However, these figures alone provide only high-level insight; a more detailed analysis can be provided by breaking down the process into units. Therefore, identifying potential KPIs requires data: both the "inputs" into LCA studies and the "output" results, with the latter ideally broken down to give an idea of contributing factors.

These KPIs can be normalized for comparison on a number of different units: volume, mass, energy content, or FU. It is advisable that volume or mass is only used when comparing chemically identical products. If the products are not chemically identical, comparing energy content is a more reliable metric. An alternative recommended approach is to compare on a functional basis (e.g., 100 km driven in a specified vehicle), with this being of particular use for assessments, with a specific functional purpose in mind.

6.4.1 Fermentation

On a cradle-to-gate basis, two major stages can be identified:

1. Cultivation stage: inputs and impacts associated with biomass cultivation
2. Production stage: inputs and impacts associated with biochemical conversion

Muñoz et al. (2014), Rathnayake et al. (2018), and Wang et al. (2013) provided similar inputs lists for cultivation, covering fertilizer, seed material, pesticide/herbicides, fuel (diesel), water, and transport. For production, the following list from the prior sources is amalgamated: feedstock, heat, electricity, chemical additives, enzymes, and water. For production, "chemical additives" and "enzymes" are quoted, as each process route requires differing inputs, many of which can contribute significantly to the overall environmental impact.

Table 6.7 gives several suggested input KPIs to consider. These have been selected due to their significant contributions to the overall environmental impact.

Although comparisons between studies can be difficult, comparisons made within studies can demonstrate the impact of varying technology choices, feedstocks, and locations. One example was provided by Muñoz et al. (2014), with the results transposed into Table 6.8, where a range of locations and feedstocks are compared on six impact categories. All cases are shown to perform better than the reference process (production from natural gas in Europe) in terms of GWP. However, some of them perform worse in terms of other environmental impact categories.

Table 6.7 Suggested KPIs for comparison of studies.

KPI	Unit
Cultivation	
N fertilizer	kg N/ha
P fertilizer	kg P_2O_5/ha
K fertilizer	kg K_2O/ha
Diesel fuel	kg/ha
Pesticide and herbicide	kg/ha
Production	
Feedstock	kg/unit product
Electricity	kWh/unit product
Heat	MJ/unit product
Chemical additives	kg/unit product

A second example is included in Table 6.9, where the method of producing bioethanol is the same in all five cases, except the pretreatment. Although the spread in results is lower than in Table 6.8, it can be seen that the pretreatment itself can have a significant impact on the overall environmental impact.

6.4.2 Anaerobic digestion

The broad range of AD feedstocks makes it hard to compare any "cultivation" phase directly. Many feedstocks are wastes (e.g., sewage sludge, agricultural residues), and once again, there is an issue of allocation. As with fermentation, pretreatment and product separation can affect the overall environmental impact. The FUs for LCAs for AD plants vary greatly. Common examples include MJ of biogas, MJ of biomethane, kWh of electricity (often used when biogas is used for power generation), and kg of feedstock (Bacenetti et al., 2016).

A number of studies were assessed to help in determining KPIs for AD (Ardolino et al., 2018; Cremiato et al., 2018; Bacenetti et al., 2016), with these summarized in Table 6.10. Of most importance, here are the energy flows. Chemicals are added to digesters for various reasons (e.g., pH maintenance), but their consumption rate is relatively low in general.

Table 6.8 Environmental impacts associated to the production of bioethanol from different agricultural feedstocks.

Impact category	Unit (per kg ethanol)	Sugarcane, Brazil (Center-South)	Sugarcane, Brazil (North-East)	Maize grain, USA	Maize stover, USA	Sugar beet, France	Wheat, France	Fossil/ Reference, Europe
Global warming potential	kg CO_2-eq	1.60	1.61	1.60	1.25	1.27	2.07	3.74
Photochemical oxidant formation potential	kg NMVOC	0.029	0.029	0.003	0.003	0.005	0.004	0.006
Terrestrial acidification potential	kg SO_2-equiv.	0.012	0.011	0.009	0.009	0.007	0.015	0.003
Freshwater eutrophication potential	kg P-equiv.	1.3×10^{-4}	1.4×10^{-4}	4.5×10^{-4}	3.9×10^{-4}	1.8×10^{-4}	2.1×10^{-4}	4.7×10^{-4}
Marine eutrophication potential	kg N-equiv.	0.0130	0.0120	0.0060	0.0040	0.0020	0.0170	0.0001
Agricultural land occupation	m^2 year	2.07	2.95	1.23	0.93	0.87	2.78	0.01

Source: Comparison data from Muñoz, I., Flury, K., Jungbluth, N., Rigarlsford, G., i Canals, L.M. and King, H., 2014. Life cycle assessment of bio-based ethanol produced from different agricultural feedstocks. The International Journal of Life Cycle Assessment 19 (1), 109–119 (Muñoz et al., 2014).

Table 6.9 Environmental impacts associated to the production of bioethanol with different pre-treatments. FU is the amount of ethanol needed to drive 1 km in a "flexible-fuel vehicle."

Impact category	Unit per FU	Pretreatment method (all pathways utilize wheat straw to produce bioethanol)				
		Dilute acid	Steam explosion with H_2SO_4 catalyst	Steam explosion without catalyst	Liquid hot water	Wet oxidation
Abiotic depletion	kg Sb-eq	1.19×10^{-3}	1.13×10^{-3}	3.96×10^{-4}	6.85×10^{-4}	4.10×10^{-4}
Global warming potential, 100 years	kg CO_2-eq	0.264	0.212	0.134	0.156	0.166
Acidification potential	kg SO_2-eq	2.18×10^{-3}	1.68×10^{-3}	9.63×10^{-4}	1.09×10^{-3}	9.67×10^{-4}
Eutrophication potential	kg PO_4-eq	4.33×10^{-4}	3.03×10^{-4}	2.12×10^{-4}	2.33×10^{-4}	2.02×10^{-4}
Ozone depletion potential	kg CFC^{-11}-eq	2.35×10^{-8}	1.68×10^{-8}	1.19×10^{-8}	1.29×10^{-8}	1.28×10^{-8}
Photochemical ozone depletion potential	kg C_2H_4-eq	1.27×10^{-4}	1.07×10^{-4}	7.09×10^{-5}	6.88×10^{-5}	7.14×10^{-5}
Human toxicity potential	kg 1,4-DB-eq	1.09×10^{-1}	8.33×10^{-2}	3.36×10^{-2}	4.34×10^{-2}	3.76×10^{-2}
Freshwater aquatic eco-toxicity potential	kg 1,4-DB-eq	4.22×10^{-2}	3.72×10^{-2}	2.49×10^{-3}	7.47×10^{-3}	5.68×10^{-3}
Terrestrial eco-toxicity potential	kg 1,4-DB-eq	1.42×10^{-3}	9.36×10^{-4}	5.81×10^{-6}	1.17×10^{-4}	5.56×10^{-5}

Source: Comparison data from Wang, L., Littlewood, J. and Murphy, R.J., 2013. Environmental sustainability of bioethanol production from wheat straw in the UK. Renewable and Sustainable Energy Reviews 28, 715–725 (Wang et al., 2013).

Table 6.10 KPIs for AD.

KPI	Unit
Feedstock input	kg/unit product
Electricity consumption	kWh/unit product
Biochemical oxygen demand	mg O_2/L
Chemical oxygen demand	mg O_2/L
Pollutants in discharge water	mg/unit product
Heat consumption	MJ/unit product
Biomethane generated	m^3/unit product
Digestate produced	kg/unit product

6.5 Product environmental footprint

In order to support the use of LCA to assess the environmental impacts of products, the Joint Research Centre (JRC) of the European Commission proposed the PEF. PEF provides the detailed requirements needed to measure and communicate the environmental impact of products taking into consideration of their entire life cycle. A PEF assessment follows the LCA methodology but provides more specific requirements for individual product categories. Therefore, more sound comparisons of products within the same category are possible, and results are more reproducible.

The PEF method was first defined and described in the EU Recommendation 2013/179/EU (European Commission, 2013). It is based on the international life cycle data (ILCD) system, also developed by the JRC. The development and implementation of the PEF method have been divided into several phases (European Platform on Life Cycle Assessment, 2021). The preparatory phase, between 2008 and 2013, was used to define the product environmental footprint category rules (PEFCRs). PEFCRs provide detailed technical guidance to perform a PEF assessment to a specific category of products, making it easier to compare the environmental performance of products within a product category. The pilot phase lasted between 2013 and 2018. It was used to test the process for developing a product, sector-specific rules, different approaches to verification, and how to communicate results to stakeholders. The transition phase started in 2019 and is currently ongoing. It aims at further improving the methodology and providing a framework for monitoring the implementation of existing PEFCRs as well as developing new PEFCRs. The last phase will be used to determine when PEF regulations are mandatory. A similar method, the Organisation

Environmental Footprint has been developed in parallel to assess the environmental impact of organizations.

The JRC of the European Commission recently provided recommendations to update the PEF method (Zampori and Pant, 2019) and develop

Table 6.11 Indicators for the PEF LCIA method (Fazio et al., 2018).

Impact category	Indicator	Unit
Climate change	Radiative forcing as Global Warming Potential (GWP100)	kg CO_2-eq
Ozone depletion	Ozone Depletion Potential (ODP)	kg CFC-11-eq
Human toxicity, cancer effects	Comparative Toxic Unit for humans (CTUh)	CTUh
Human toxicity, noncancer effects	CTUh	CTUh
Particulate matter/ Respiratory inorganics	Human health effects associated with exposure to $PM_{2.5}$	Disease incidences
Ionizing radiation, human health	Human exposure efficiency relative to U^{235}	kBq U^{235}
Photochemical ozone formation	Tropospheric ozone concentration increase	kg NMVOC-eq
Acidification	Accumulated exceedance (AE)	mol H + -eq
Eutrophication, terrestrial	AE	mol N-eq
Eutrophication, aquatic freshwater	Fraction of nutrients reaching freshwater end compartment (P)	kg P-eq
Eutrophication, aquatic marine	Fraction of nutrients reaching marine end compartment (N)	kg N-eq
Eco-toxicity (freshwater)	Comparative toxic unit for ecosystems (CTUe)	CTUe
Land use	Soil quality index (biotic production, erosion resistance, mechanical filtration, and groundwater replenishment)	Dimensionless
Water scarcity	User deprivation potential (deprivation-weighted water consumption)	kg water-eq deprived
Resource use, minerals, and metals	Abiotic resource depletion (ADP ultimate reserves)	kg Sb-eq
Resource use, energy carriers	Abiotic resource depletion—fossil fuels (ADP—fossil)	MJ

process data sets compliant with the environmental footprint (EF) requirements (Fazio et al., 2020). A detailed guidance document was also prepared to explain how to perform a PEF assessment and develop PEFCRs (Manfredi et al., 2012).

The PEF method recommends following the EF life cycle impact assessment method. In this method, only midpoint indicators are used. The impact categories, indicators, and units used in the PEF method can be seen in Table 6.11. Fazio et al. (2018) described the characterization factors and methods used to calculate the LCIA indicators.

It is also essential to consider how the information obtained with a PEF analysis should be used and communicated. Elsen et al. (2019) described how to communicate PEF information effectively to consumers to make them more aware of the environmental consequences of their purchasing decisions. Lupiáñez-Villanueva et al. (2018) recommended displaying an A—E rating and an average score on the product label. It is expected that the PEF will raise public awareness on the importance of considering the footprint of products as well as encourage manufacturers to minimize the environmental impact of their products.

A number of PEFCRs have been developed for various products, including different food products, IT equipment, decorative paints, metal sheets, etc. Some examples of the use of PEFCRs to analyze the environmental impact of products include the analysis of dairy products and photovoltaic modules (Egas et al., 2020; Wade et al., 2018). Unfortunately, no PEFCRs have been developed yet for biofuels or, more generally, for fuels. It is encouraged to develop such PEFCRs for fuels so more systematic, reliable, and reproducible studies could be performed to assess their environmental impact. This is particularly important in the case of biofuels, as these PEFCRs would allow comparing biofuels with conventional fuels more accurately and confirm whether biofuels generate a lower environmental impact than conventional fuels.

6.6 Conclusions

Biogas and bioethanol are the most common biofuels produced from biomass. They are mostly produced via AD and fermentation, respectively. Several steps are involved in such biochemical conversions, each influencing the yield and overall process performance. Furthermore, each step generates an environmental impact, which should be aggregated and considered along the environmental impacts created in all life cycle stages of the product. The best way to systematically account for these environmental impacts is by using the LCA methodology. This methodology enables fair comparisons between biofuels and conventional fuels, considering a wide range of impact categories, such as GWP, abiotic depletion, human toxicity potential, and land use. In this way, it can be ascertained whether biofuels do perform better environmentally than their fossil-fuel counterpart. The PEF provides more specific requirements for performing an LCA for individual product categories. Currently, there are no PEF category rules for biofuels nor fuels in general, and their development to support LCA studies of biofuels is highly encouraged.

Acknowledgements

Guillermo Garcia-Garcia acknowledges the Grant Juan de la Cierva Incorporación funded by MCIN/AEI/10.13039/501100011033 by "ESF Investing in your future".

References

Abdullah, B., Muhammad, S.A.F.A.S., Shokravi, Z., Ismail, S., Kassim, K.A., Mahmood, A.N., et al., 2019. Fourth generation biofuel: a review on risks and mitigation strategies. Renewable and Sustainable Energy Reviews 107, 37–50.

Abo, B.O., Gao, M., Wang, Y., Wu, C., Wang, Q., Ma, H., 2019. Production of butanol from biomass: recent advances and future prospects. Environmental Science and Pollution Research 26 (20), 20164–20182.

Abraham, A., Mathew, A.K., Park, H., Choi, O., Sindhu, R., Parameswaran, B., et al., 2020. Pretreatment strategies for enhanced biogas production from lignocellulosic biomass. Bioresource Technology 301, 122725.

Ardolino, F., Parrillo, F., Arena, U., 2018. Biowaste-to-biomethane or biowaste-to-energy? An LCA study on anaerobic digestion of organic waste. Journal of Cleaner Production 174, 462–476.

Atelge, M.R., Atabani, A.E., Banu, J.R., Krisa, D., Kaya, M., Eskicioglu, C., et al., 2020. A critical review of pretreatment technologies to enhance anaerobic digestion and energy recovery. Fuel 270, 117494.

Bacenetti, J., Sala, C., Fusi, A., Fiala, M., 2016. Agricultural anaerobic digestion plants: what LCA studies pointed out and what can be done to make them more environmentally sustainable. Applied Energy 179, 669–686.

Balan, V., 2014. Current challenges in commercially producing biofuels from lignocellulosic biomass. International Scholarly Research Notices 2014, 463074.

Barros, S., 2019. Brazil Biofuels Annual 2019, Global Agricultural Information Network (GAIN) Report. <https://apps.fas.usda.gov/newgainapi/api/report/downloadreportbyfilename?filename = Biofuels%20Annual_Sao%20Paulo%20ATO_Brazil_8-9-2019.pdf> (accessed May 25, 2021).

Bedoić, R., Čuček, L., Ćosić, B., Krajnc, D., Smoljanić, G., Kravanja, Z., et al., 2019. Green biomass to biogas—a study on anaerobic digestion of residue grass. Journal of Cleaner Production 213, 700–709.

BP, 2020. Statistical Review of World Energy, 69th ed. <https://www.bp.com/content/dam/bp/business-sites/en/global/corporate/pdfs/energy-economics/statistical-review/bp-stats-review-2020-full-report.pdf> (accessed May 25, 2021).

Chen, Q., Wu, W., Qi, D., Ding, Y., Zhao, Z., 2020. Review on microaeration-based anaerobic digestion: state of the art, challenges, and prospectives. Science of the Total Environment 710, 136388.

Chum, H., Faaij, A., Moreira, J., Berndes, G., Dhamija, P., Dong, H., et al., 2011. In: Edenhofer, O., Pichs-Madruga, R., Sokona, Y., Seyboth, K., Matschoss, P., Kadner, S., Zwickel, T., Eickemeier, P., Hansen, G., Schlömer, S., et al. (Eds.), IPCC Special Report on Renewable Energy Sources and Climate Change Mitigation, pp. 46–59.

Cremiato, R., Mastellone, M.M., Tagliaferri, C., Zaccariello, L., Lettieri, P., 2018. Environmental impact of municipal solid waste management using Life Cycle Assessment : The effect of anaerobic digestion, materials recovery and secondary fuels production. Renewable Energy 124, 180–188.

Egas, D., Ponsá, S., Colon, J., 2020. CalcPEFDairy: a Product Environmental Footprint compliant tool for a tailored assessment of raw milk and dairy products. Journal of Environmental Management 260, 110049.

Elsen, M., van Giesen, R., van den Akker, K., Dunne, A., 2019. Consumer testing of alternatives for communicating the Environmental Footprint profile of products. Final report, European Commission.

European Commission (EC), 2013. Commission Recommendation of 9 April 2013 on the use of common methods to measure and communicate the life cycle environmental performance of products and organisations. ANNEX II: Product Environmental Footprint (PEF) Guide. European Commission.

European Platform on Life Cycle Assessment, 2021. Environmental Footprint. <https://eplca.jrc.ec.europa.eu/EnvironmentalFootprint.html> (accessed May 05, 2021).

Fazio, S., Castellani, V., Sala, S., Schau, E.M., Secchi, M., Zampori, L., 2018. Supporting information to the characterisation factors of recommended EF Life Cycle Impact Assessment methods. EUR 28888 EN, European Commission, Ispra, ISBN 978-92-79-76742-5. doi: 10.2760/671368, JRC109369.

Fazio, S., Zampori, L., De Schryver, A., Kusche, O., Thellier, L., Diaconu, E., 2020. Guide for EF compliant data sets, Version 2.0 Luxembourg, ISBN 978-92-76-17951-1 (online). doi:10.2760/537292 (online), JRC120340.

Fokaides, P.A., Christoforou, E., 2016. In: Luque, R., Sze Ki Lin, C., Wilson, K., Clark, J. (Eds.), Handbook of Biofuels Production, second ed. Woodhead Publishing Ltd, United Kingdom.

Houfani, A.A., Anders, N., Spiess, A.C., Baldrian, P., Benallaoua, S., 2020. Insights from enzymatic degradation of cellulose and hemicellulose to fermentable sugars—a review. Biomass and Bioenergy 134, 105481.

International Energy Agency (IEA), 2020. Renewables 2020. <https://www.iea.org/reports/renewables-2020> (accessed April 21, 2021).

International Standard Organisation (ISO 14040), 2006. Environmental Management - Life Cycle Assessment - Principles and Framework. <https://www.iso.org/standard/37456.html> (accessed April 01, 2021).

Internatioanl Standard Organisation (ISO 14044), 2006. Preview Environmental Management—Life Cycle Assessment—Requirements and *Guidelines*. <https://www.iso.org/standard/38498.html> (accessed April 01, 2021).

Jain, S., Newman, D., Nzihou, A., Dekker, H., Le Feuvre, P., Richter, H., et al., 2019. Global Potential of Biogas (Doctoral dissertation). The World Biogas Association.

Jensen, I.G., Münster, M., Pisinger, D., 2017. Optimizing the supply chain of biomass and biogas for a single plant considering mass and energy losses. European Journal of Operational Research 262 (2), 744–758.

Liang, F., Englund, E., Lindberg, P., Lindblad, P., 2018. Engineered cyanobacteria with enhanced growth show increased ethanol production and higher biofuel to biomass ratio. Metabolic Engineering 46, 51–59.

Lupiáñez-Villanueva, F., Tornese, P., Veltri G.A., Gaskell, G., 2018. Assessment of different communication vehicles for providing Environmental Footprint information. Request for Specific Services for the Implementation of the Framework Contract no. EAHC-2011-CP-01. Final Report. <https://ec.europa.eu/environment/eussd/smgp/pdf/2018_pilotphase_commreport.pdf> (accessed May 11, 2021).

Manfredi, S., Allacker, K., Pelletier, N., Chomkhamsri, K., de Souza, D.M., 2012. Product environmental footprint (PEF) guide.

Manochio, C., Andrade, B.R., Rodriguez, R.P., Moraes, B.S., 2017. Ethanol from biomass: a comparative overview. Renewable and Sustainable Energy Reviews 80, 743–755.

Moreno, A.D., Alvira, P., Ibarra, D., Tomás-Pejó, E., 2017. Production of ethanol from lignocellulosic biomass. Production of Platform Chemicals from Sustainable Resources. Springer, Singapore, pp. 375–410.

Muñoz, I., Flury, K., Jungbluth, N., Rigarlsford, G., i Canals, L.M., King, H., 2014. Life cycle assessment of bio-based ethanol produced from different agricultural feedstocks. The International Journal of Life Cycle Assessment 19 (1), 109–119.

National Renewable Energy Laboratory, 2015. U.S. Life Cycle Inventory Database. https://www.lcacommons.gov (accessed 11 April 2022).

Pandiyan, K., Singh, A., Singh, S., Saxena, A.K., Nain, L., 2019. Technological interventions for utilization of crop residues and weedy biomass for second generation bioethanol production. Renewable Energy 132, 723–741.

Rasapoor, M., Young, B., Brar, R., Sarmah, A., Zhuang, W.Q., Baroutian, S., 2020. Recognizing the challenges of anaerobic digestion: critical steps toward improving biogas generation. Fuel 261, 116497.

Rastogi, M., Shrivastava, S., 2017. Recent advances in second generation bioethanol production: an insight to pretreatment, saccharification and fermentation processes. Renewable and Sustainable Energy Reviews 80, 330–340.

Rathnayake, M., Chaireongsirikul, T., Svangariyaskul, A., Lawtrakul, L., Toochinda, P., 2018. Process simulation based life cycle assessment for bioethanol production from cassava, cane molasses, and rice straw. Journal of Cleaner Production 190, 24–35.

RFA, 2020. Focus Forward: 2020 Ethanol Industry Outlook. <https://ethanolrfa.org/file/1537/2020-Outlook-Final-for-Website.pdf> (accessed May 23, 2021).

Rizza, L.S., Coronel, C.D., Smachetti, M.E.S., Do Nascimento, M., Curatti, L., 2019. A semi-closed loop microalgal biomass production-platform for ethanol from renewable sources of nitrogen and phosphorous. Journal of Cleaner Production 219, 217–224.

Roj-Rojewski, S., Wysocka-Czubaszek, A., Czubaszek, R., Kamocki, A., Banaszuk, P., 2019. Anaerobic digestion of wetland biomass from conservation management for biogas production. Biomass and Bioenergy 122, 126–132.

Susmozas, A., Martín-Sampedro, R., Ibarra, D., Eugenio, M.E., Iglesias, R., Manzanares, P., et al., 2020. Process strategies for the transition of 1G to advanced bioethanol production. Processes 8 (10), 1310.

Tsapekos, P., Kougias, P.G., Treu, L., Campanaro, S., Angelidaki, I., 2017. Process performance and comparative metagenomic analysis during co-digestion of manure and lignocellulosic biomass for biogas production. Applied Energy 185, 126–135.

Veza, I., Said, M.F.M., Latiff, Z.A., 2021. Recent advances in butanol production by acetone-butanol-ethanol (ABE) fermentation. Biomass and Bioenergy 144, 105919.

Wade, A., Stolz, P., Frischknecht, R., Heath, G., Sinha, P., 2018. The Product Environmental Footprint (PEF) of photovoltaic modules—Lessons learned from the environmental footprint pilot phase on the way to a single market for green products in the European Union. Progress in Photovoltaics: Research and Applications 26 (8), 553–564.

Wang, L., Littlewood, J., Murphy, R.J., 2013. Environmental sustainability of bioethanol production from wheat straw in the UK. Renewable and Sustainable Energy Reviews 28, 715–725.

Zabed, H.M., Akter, S., Yun, J., Zhang, G., Zhang, Y., Qi, X., 2020. Biogas from microalgae: technologies, challenges and opportunities. Renewable and Sustainable Energy Reviews 117, 109503.

Zampori, L., Pant, R., 2019. Suggestions for updating the Product Environmental Footprint (PEF) method. JRC Technical Repports. Publications Office of the European Union, Luxembourg, p. 76.

Zhao, T., Tashiro, Y., Sonomoto, K., 2019. Smart fermentation engineering for butanol production: designed biomass and consolidated bioprocessing systems. Applied Microbiology and Biotechnology 103 (23), 9359–9371.

CHAPTER 7

Environmental assessment of biomass-to-biofuels mechanical conversion routes (pelleting, briquetting)

Elias Christoforou
School of Engineering and Applied Sciences, Frederick University, Nicosia, Cyprus

Contents

7.1 Introduction

Bioenergy and consequently the production of sustainable biofuels can play a significant role in achieving the global targets for greenhouse gas (GHG) mitigation and sustainable development. Currently, bioenergy is the main source of renewable energy in the European Union (EU), accounting for more than 63% of the renewable energy consumption (Bioenergy Europe, 2019). The share of bioenergy in the global energy mix is expected to increase in the following years, aiming at the challenging target of at least 60% of global primary energy and 75% of electricity coming from zero-carbon technologies and options by 2050.

The higher contribution of bioenergy on total global energy consumption, with the heat sector being the primary market for biofuels,

Environmental Assessment of Renewable Energy Conversion Technologies
DOI: https://doi.org/10.1016/B978-0-12-817111-0.00009-7

simultaneously leads to an increase of the global market for solid biofuels, which currently counts for more than 69% of the gross inland energy consumption in EU28 (Bioenergy Europe, 2019).

Scientific research focuses the last years not only on obtaining high-quality solid biofuels but also on identifying the most sustainable production pathways, taking into consideration the three pillars of sustainability: economic growth, environmental protection, and social equity (Fokaides and Christoforou, 2016). Pelleting and briquetting are two widely applied pre-treatment technologies for solid biomass feedstock and the most common technologies for biomass densification. The technologies allow the transformation and energy upgrading of solid biomass feedstock into a valuable biofuel with superior thermochemical and combustion characteristics as well as advanced handling, storage, and transportation properties (Ruiz et al., 2018). It is worth noting that among the consumed solid biomass in EU28 in 2017, pellets shared approximately 10% of biomass energy, with the global pellet market showing an increasing trend (Fig. 7.1). Several studies have been conducted in the last decade trying to estimate the environmental impact of those pretreatment technologies. Among several methods, life cycle assessment (LCA), a well-established and standardized method for assessing the sustainability of a product, process, or service over its entire life cycle, has been most widely used. This chapter aims to present the current status of LCA of biomass pelleting and briquetting. Section 7.2 presents the main processes related to two pretreatment methods, followed by

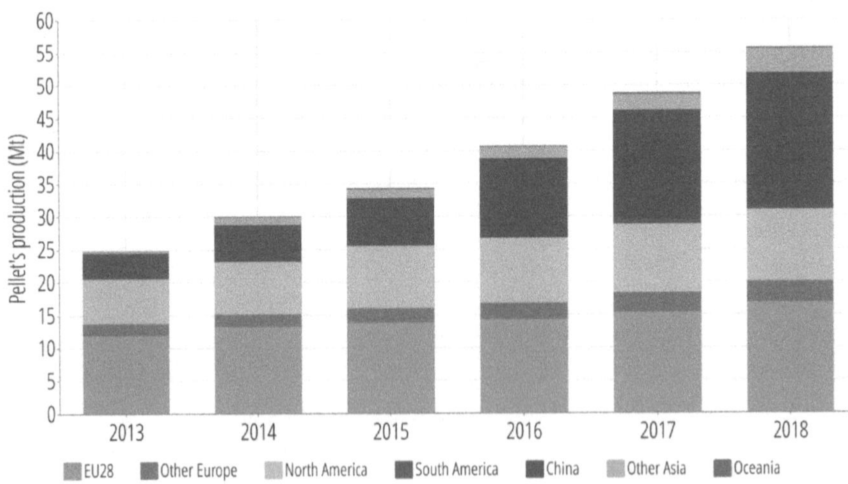

Figure 7.1 World pellet production in Mt. *Data from Bioenergy Europe, 2019.*

an introduction to the LCA methodology and framework in Section 7.3. The results of previous studies in this specific field are presented and discussed in Section 7.4.

7.2 Pelleting and briquetting

Pelleting and briquetting are two widely applied methods for upgrading solid biomass feedstock, leading to the formation of designated homogeneous, high-quality solid biofuels. The processed raw material, usually in fine particle form, is converted to a compacted end-product with advanced physical and mechanical properties, lower moisture content, and higher energy density. The loose raw feedstock is converted under high-pressure and high-temperature conditions, due to the presence of friction between the surfaces of the processing equipment and the feedstock particles, into a valuable end-product (Klemm et al., 2013).

Of course, several parameters affect the quality and the properties of the end-product, including, but not limited to, the particle size of the raw material, its initial moisture content, and its chemical composition. The equipment and processing parameters, such as the temperature and compression pressure, also play a significant role in the derived product (Kirsten et al., 2016).

7.2.1 Pelleting/briquetting feed biomass

In literature, several biomass feedstocks have been used for pelleting and briquetting. Lignocellulosic biomass feedstocks (i.e., agricultural and wood residues) have received the most significant attention as raw materials for densification; since the final products, pellets or briquettes, are considered carbon-neutral (Shahrukh et al., 2015). Nowadays, woody biomass is the most widely used feedstock for pelleting, while agricultural feedstock is mainly utilized as feedstock for the production of briquettes.

Raw woody biomass, as well as the main by-products and wastes of forest management and woodworking/processing industry (wood chips, wood shavings, and sawdust), have been extensively processed, either for pelleting or briquetting, based on the different characteristics of the feedstock (Ruiz et al., 2018). An alternative biomass-to-biofuel stream that receives significant attention over the last decade, is the cultivation—in a sustainable manner—of fast-growing species as energy crops (Fantozzi and Buratti, 2010; Ruiz et al., 2018).

7.2.2 Pellets/briquettes classification

The global debate regarding the sustainability of biofuels and the fact that the quality and availability of woody biomass is not always as expected and consistent, push both academia and industry to develop alternative biofuel production chains. These efforts will ensure feedstock availability for biomass densification in a demand-increasing environment.

The Technical Committee (i.e., TC 335), developed by the European Committee for Standardization, carried out significant work in the field of solid biofuels standardization, considering several types of biomass feedstocks to be used as a source for solid biofuels.

As for solid biofuels in the form of pellets and briquettes, the following classification has been proposed and established:

- WPs for supply to large-scale production of heat and power.
- Packaged pellets for the residential market.
- Briquetted wood for the residential market and smaller industrial heat-producing facilities.
- Agro-pellets, used by domestic business transactions.

The first part of the EN ISO 17225 standard series distinguishes biomass into specific groups: woody, herbaceous, fruit, and aquatic biomass. A general group was also defined to cover any blend or mixtures of materials that originated from the first groups.

Regarding pellets and briquettes, specific standards were developed to define/set their minimum quality requirements.

The accepted size and required thermochemical and mechanical properties of woody and nonwoody pellets and briquettes are well defined in the below standards:

- EN ISO 17225–2: fuels specification and classes—part 2: graded WPs.
- EN ISO 17225–3: fuels specification and classes—part 3: graded wood briquettes.
- EN ISO 17225–6: fuels specification and classes—part 6: graded nonwoody pellets.
- EN ISO 17225–7: fuels specification and classes—part 7: graded nonwoody briquettes.

7.2.3 Process description

Pelleting and briquetting technologies present similar process stages; however, they differ on the type of equipment used during the compaction of

Figure 7.2 Production process of pellets and briquettes.

the material as well as the size and shape of the delivered products. The typical pelleting/briquetting system consists of the below stages (Fig. 7.2):

- Feedstock/raw materials collection: the feedstock to be processed is collected at the pretreatment plant and stored in predefined places (e.g., storage rooms, silos). Due to the raw material's low bulk density and the need to keep the transportation costs as low as possible, pelleting and briquetting are applied to locally available feedstock.
- Drying: this process is necessary for both pelleting and briquetting systems to obtain a homogeneous—in terms of moisture content—processed feedstock. This parameter is among the most important variables that affect the binding mechanisms and therefore, the quality of the product. In general, a moisture content of the feedstock of around 12% is desired when a final product with a moisture content of 6%–8% is targeted (Karkania et al., 2012).
- Particle size reduction/grinding/milling: the dried feedstock is usually ground to obtain a uniform particle size of the feedstock (typically to a size lower than the desired size of the pellet). Grinding enables faster drying of loose material when fed for densification. The desired size reduction percentage is based on the pelleting/briquetting equipment to be used.
- Mixing: this stage involves, where applicable, intentional mixing of the processed feedstock or blends of feedstocks with additives (e.g., sawdust). This aims to achieve better product quality or making the process more efficient.

- Conditioning: at this stage, if necessary, conditioning and softening of the mixture are taking place, usually by using superheated stream or injection of water prior to its densification.
- Pelletizing/briquetting: the preconditioned, homogeneous feedstock is fed to the pelletizing/briquetting machines where densification occurs. Pellets are mainly produced via roller press machines. The two main types of conventional pelletizers are the flat die roller and ring die press. In a flat die roller press, rollers rotate on a stationary circular perforated disk, compressing the processed feedstock that is usually fed from the top of the pellet machine. Pellets produced in a flat die roller press are usually homogeneous due to uniform extrusion and even material densification. On a ring-type press, rollers remain stationary while the perforated disk rotates and presses the feedstock onto the inner perimeter (Chen et al., 2009). Briquettes are usually produced using hydraulic, mechanical, or roller presses. Piston-pressed briquetting involves punching the feedstock into a die with a reciprocating ram or a plunger to obtain a briquette. The movement of the piston is mechanically achieved or by means of a hydraulic transmission. Screw presses, on the other hand, extrude the end-product employing single- or double-screw extruders through an externally heated taper die (Chen et al., 2009).
- Cooling: this stage is mainly applied in pellet production systems and significantly affects the mechanical properties of the pellets, specifically their strength and durability. Air is supplied on the production line in order to cool the hot pellets collected from the pelleting machine, enabling the product to become more robust and solid.
- Screening/sieving: during this stage, fine material is removed from the cooled pellets by using mechanical vibration/screening, enabling their packaging and preparation for transportation.
- Packaging and storing: the solid biofuels are packaged and stored until their transportation to end-users.

7.3 Life cycle assessment

LCA is a well-structured and comprehensive method for assessing the environmental impact of a product or system over its entire life cycle from the extraction of raw material and other resources, the production and use phase, up to the disposal of waste material. It is a valuable tool for the estimation of all relevant inputs and outputs, as well as the environmental footprint on specific impact categories of different processes within an investigated system (Kylili et al., 2016).

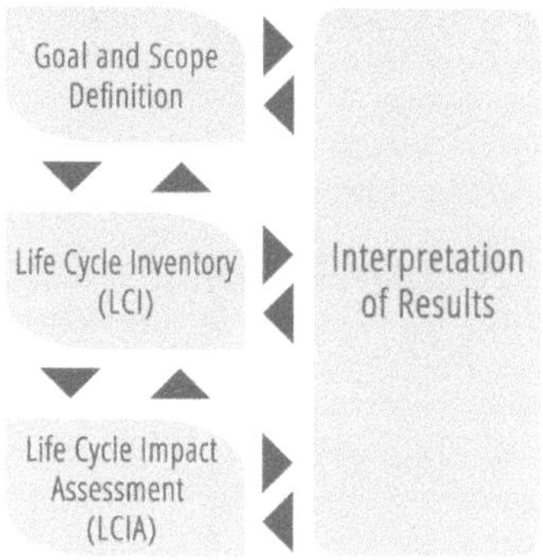

Figure 7.3 Life cycle assessment stages (Reproduced with permission from Fokaides and Christoforou, 2016).

LCA methodological approach is standardized as per ISO 14040 (ISO, 2006) and is distinguished into four steps (Fig. 7.3):

- Goal and scope definition
- Life cycle inventory (LCI)
- Life cycle impact assessment (LCIA)
- Interpretation of results.

LCA studies vary in terms of their predefined scope and goal, system boundaries, reference system or functional unit, or applied allocation method (e.g., attributional or consequential). For example, an attributional LCA describes the environmentally relevant flows to and from a life cycle, and its subsystems and coproduction are treated by applying allocation factors. On the other hand, a consequential LCA describes how changes in environmental relevant flows will change in response to possible decisions in the future (Fokaides and Christoforou, 2016). Therefore, the comparison of results derived from different studies always needs to be undertaken with care.

7.3.1 General life cycle assessment framework

7.3.1.1 Goal and scope definition

The first stage of an LCA study can be considered as the most important, since, at this point, the study's goal and scope are clearly stated, the boundaries of the system under consideration are defined, and the appropriate functional unit, as well as the reference system, are selected. Also, at this point, any assumptions, limitations, or data quality requirements are defined.

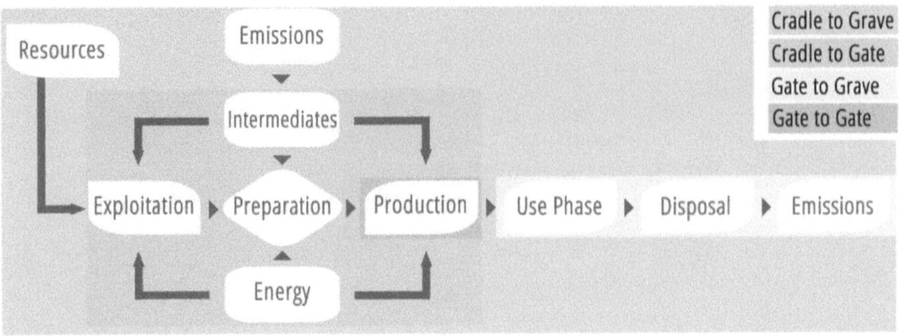

Figure 7.4 Life cycle assessment-system boundaries (Reproduced with permission from Fokaides and Christoforou, 2016).

- Goal and scope definition: the goal definition of the study will specify the objective of the LCA study as well as the application of the results, while through the definition of scope, the main characteristics and the geographical extent of the study will be justified.
- Functional unit: the selected functional unit (FU) is considered a reference point at which all the inputs and outputs within the boundaries of the system are related.
- System boundaries: the definition of system boundaries will indicate exactly which processes within the system's entire life cycle will be taken under consideration for the study. Below, different approaches related to biofuels LCA are defined (Fig. 7.4):
  - Cradle-to-gate: this approach excludes the use and disposal phase of the system's life cycle and focuses on the preparation and production phase of a product.
  - Cradle-to-grave: this approach examines the entire life cycle of a product and takes into consideration all the inputs and outputs from processes during raw material extraction, production, transportation to the user, use phase, and waste disposal.
  - Gate-to-grave: this analysis excludes all the preparatory processes and focuses on the use and end-of-life phases of a product once it leaves the factory.
  - Gate-to-gate: the environmental impact of a specific process in the system is analyzed.
- Reference system: the selected reference system for comparison of the study results shall have the same critical characteristics as the system under investigation, namely a common FU, similar system boundaries as well as similar geographical and temporal context.

7.3.1.2 Life cycle inventory

LCI is of critical importance since it involves:

- The collection of necessary data for the definition of all unit processes that will be included in the system boundaries.
- Performingallocation steps in case of multifunctional processes, in order to relate all inputs (materials, energy resources) and outputs (emissions, end-products, wastes, etc.) with the relevant products and by-products, if applicable.
- Final calculations with validation of collected data and application of allocation procedures.

Existing commercial and public databases such as Ecoinvent (Ecoinvent Centre, 2017), Biograce (IEE, 2011), US Life Cycle Inventory, Database (US Life Cycle Inventory Database, 2012), and the European Life Cycle Database (ELCD, 2006) can significantly simplify the data collection process, provided that the required quality of data is maintained. As per ISO 14044 (ISO, 2006b), the quality of collected data can be addressed based on the below key categories:

- Time-related coverage
- Geographical coverage
- Technology coverage
- Precision
- Completeness
- Representativeness
- Consistency
- Reproducibility
- Data sources
- Uncertainty of data.

7.3.1.3 Life cycle impact assessment

Based on the outcome of LCI, this stage evaluates the environmental impact of the investigated system under specific impact categories. As per ISO 14044 (ISO, 2006b), the below mandatory stages shall be included in LCIA:

- Specific impact categories indicates, and characterization models shall be selected and identified.
- Sorting and combination of LCI results based on the selected impact categories.
- calculation of results per selected impact category.

Additionally, this phase may also include optional stages such as normalization, grouping, and weighting.

The normalization step aims to quantify the impact of a certain flow in the system relative to a reference value. Grouping aims to sort and possibly rank the selected impact categories into sets to better allow the interpretation of the results into specific areas of concern. Finally, weighting is when the results under different impact categories are converted and possibly aggregated using numerical factors based on value choices (ISO, 2006b).

The existing LCIA methodologies are classified into two main categories (Fokaides and Christoforou, 2016):

- Midpoint approach: the environmental impacts are expressed as environmental concerns such as climate change, acidification, human toxicity, etc.
- Endpoint approach: the environmental impacts are translated into social and other concerns like human health, natural environment, and natural resources.

7.3.1.4 Interpretation of results

The last step in an LCA study refers to the interpretation of the impact assessment results. This phase includes the completeness, sensitivity, and consistency checks:

- The completeness check is necessary to ensure the completeness of the study and that all the relevant information needed for the interpretation is available.
- Sensitivity checks aim to determine the influence that a potential change in the applied assumptions, used data, and applied LCIA methodology may have on the result.
- Finally, the consistency check aims to define whether the whole applied procedure, including data, models, assumptions, and methods, is consistent with the objectives and scope of the study.

7.3.2 Life cycle assessment components in biomass densification systems

7.3.2.1 System boundaries

As previously mentioned, the quality of LCA studies and the obtained results are highly affected by the definition of system boundaries and the functional units. Also, the definition of the various subprocesses within a densification scheme is of great importance.

As shown in Fig. 7.4, different system boundaries can be defined in an LCA study. Specifically, in an LCA study of a biomass densification system, a "cradle-to-grave" LCA would include in the analysis all the steps and processes, with all relevant energy inputs and outputs, before the transportation of feedstock at the plant's gate, as illustrated in Fig. 7.5. The analysis would

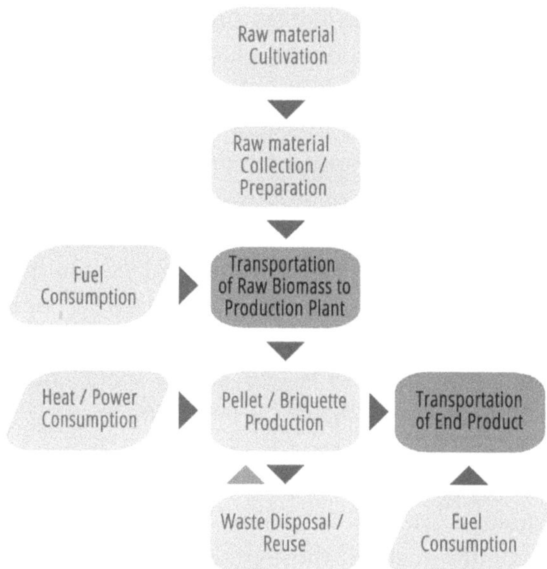

Figure 7.5 Life cycle stages considered in the implementation of life cycle assessment studies of biomass densification systems.

consider the cultivation phase, collection/preparation of biomass feedstock, and the collection of biomass residues. It would also take into consideration the disposal of wastes derived from the pellet/briquette production plant.

Accordingly, a "gate-to-gate" analysis would focus on the preparation of the feedstock, the production phase until the storing/packaging of the end-product, omitting any of the contributions of processes outside the manufacturing plant. In some cases, biomass feedstock transportation, as well as pellets/briquettes, to the distribution centers, are considered within the gate-to-gate limits. The environmental impact of the relevant preparatory (i.e., drying, milling, conditioning, mixing) and postdensification processes involved in the pellet/briquette production phase, as shown in Fig. 7.2, shall always be analyzed in a gate-to-gate system.

7.3.2.2 Functional unit

Another critical definition within a biomass densification LCA study regards the FU, since it facilitates the comparison of the obtained results with other reference systems. While the FU could be defined based on input values (e.g., the mass of processed feedstock), it is observed in the literature that most studies prefer an output related FU (e.g., mass of product delivered, unit of energy outcome from the system).

7.3.2.3 Energy inputs in densification plant subprocesses

- Storage of raw material and end-product: the process is potentially related to energy inputs, mainly the embodied energy of storage facilities (i.e., silos, warehouses), if the storage is not carried out in an open area.
- Drying process: the environmental impact of the drying process is highly dependent on the condition, in terms of moisture content of the raw material. In order to ensure high-quality densified products, the moisture percentage of the raw material should be between 10% and 20%. The total fuel consumption and the total time of operation of a biomass dryer are strictly related to the condition of the processed biomass. The drying process is one of the main contributors to the environmental impact of a densification system.
- Milling: the process is necessary for almost all pelleting systems. Milling is achieved using hammer mills or crashers, and the energy inputs in this process are strictly related to the desired percentage of size reduction.
- Mixing and conditioning: in those production systems, where multiple types of feedstocks are blended or binding materials are added to a single or multiple types of feedstocks, additional energy inputs shall be considered for the operation of a separate mixer.
- Conditioning of the feedstock may take place, if necessary, at temperatures above 100°C. The inputs in terms of energy consumption at this stage are relevant to the time the conditioning is in progress.
- Pelleting/briquetting: Energy input in this stage is affected by the capacity of the densification equipment, as well as the condition of the processed feedstock (i.e., lower energy needs for softer processed feedstock).
- Cooling/sieving: the energy inputs for this stage mainly are: (1) the operational energy consumed to supply air needed for drying the produced fresh pellets and (2) the energy needed to sieve the produced pellets and remove fine particles/dust.
- Packaging: the moderate energy inputs for packaging deal with the capacity of the packing device.

7.3.3 Previous work on life cycle assessment of biomass densification systems

This section provides an overview of results obtained from selected recent LCA studies in the field of biomass densification processes. A review of previous studies (Table 7.1) revealed that the selected FU for LCA studies of biomass densification systems varies, either defined in terms of the mass

Table 7.1 Overview of life cycle assessment (LCA) studies for biomass densification.

Study	Scope	Object(s) of investigation	Functional unit	System boundaries	Employed LCA software/databases	LCIA methodology	Sensitivity analysis
Kylili et al. (2016)	Evaluation of the environmental impacts of the pelleting process using waste olive husk for the production of pellets for domestic heat purpose.	Pellet production using olive husk—4 scenarios.	1 tonne of produced CEN compliant pellets	Cradle-to-grave	Gabi Software ANPA; Boustead Consulting	International reference life cycle data system (ILCD)	No
El Hanandeh (2015)	Management of olive oil industry waste in Australia.	• Briquette manufacturing and use for home heating in a domestic solid fuel stove. • Pellet manufacturing and use in domestic biomass water heater.	mg of solid olive waste at the mill.	Cradle-to-grave	ELCDNRELAusLCI	ReCiPe Midpoint (H) CML 2001 (for AP)	Yes
Adams et al. (2015)	LCA of the torrefaction with pelletization (TP) and wood pellet (WP) bioenergy chains on a cradle-to-gate basis.	• Compile a detailed LCI of TP/WP. • Process chains • Complete a cradle-to-gate LCA of the TP/WP bioenergy chains. • Perform an impact assessment and compare the results of the TP/WP process.	1 ton or 1 MJ of TP/WP delivered	Cradle-to-gate	SimaPro	ReCiPe	Yes

(*Continued*)

Table 7.1 (Continued)

Study	Scope	Object(s) of investigation	Functional unit	System boundaries	Employed LCA software/databases	LCIA methodology	Sensitivity analysis
		• Perform sensitivity analysis to evaluate the significance of the following key modeling assumptions:					
Quinteiro et al. (2019)	Assess the environmental impacts of five different wood-based residential heating systems in Portugal.	• A pellet stove using maritime pine pellets as feedstock. • A wood stove using eucalyptus as feedstock. • A wood stove using maritime pine split logs as feedstocks; a fireplace using eucalyptus. • A fireplace using maritime pine split logs as feedstocks.	1 MJ of thermal energy for residential heating.	Cradle-to-grave	Ecoinvent database 2017EMEP/EEA, 2017	ReCiPe 2016 midpoint	Yes
Hamedani et al. (2019)	Pellet production system from vine and olive grove woody biomass.	olive and vine pellet	1 MJ of pellet produced	cradle-to-gate	SimaPro	Recipe midpoint (H) v1.13	Yes
Laschi et al. (2016)	Evaluate the environmental impacts related to high-quality pellet production for domestic heating.	WP production—4 scenarios related to biomass combustion phase at the pellet factory.	1 kg of packed WPs.	cradle-to-gate	SimaPro	ReCiPe Midpoint v1.07	No

Ruiz et al. (2018)	Quantify the environmental performance of a biofuel chain-based WPs for use in residential heating using life cycle methodology.	Analysis of biofuel chain-based WPs considering alternative scenarios regarding plant design (conventional boiler vs modern CHP system), biomass type (pine, poplar) and origin (forest management, wood processing industry, and SRC), and the utilization of ash from the biomass boiler as a surrogate for synthetic fertilizers.	1 MJ of net energy produced by a WP domestic boiler.	Cradle-to-grave	SimaPro	ILCD	Yes

of end-product (i.e., pellet, briquette), the energy produced, or mass of the processed feedstock. Also, it was obtained that cradle-to-grave or cradle-to-gate (i.e., excluding end-product utilization) analysis is preferred, depending on the availability of data and the scope of each study.

Fig. 7.5A and B demonstrates the contribution of different life-cycle stages of densification systems and their components to the overall GHG emissions and energy consumption, respectively.

Muazu et al. (2017) conducted a comprehensive review study focused on LCA studies for densification of various feedstocks, including the olive husk, corn stover, sawdust, wood chips, etc. (Fig. 7.6). The results obtained from 19 different studies were analyzed and valuable conclusions were made. The authors noted that none of the selected studies reported a "cradle-to-grave" boundary system; thus, the production/collection of biomass feedstock, as well as the end-product use phase and waste disposal phases, were excluded from the analysis.

The review study also highlighted the fact that, although several recent LCA studies exist, focusing on biomass pelleting and briquetting production systems, only a few reported environmental impacts, specifically for the biomass densification phase.

Following a normalization step, the estimated GHG emissions retrieved for a gate-to-gate analysis were in the range of 0.6–50 CO_2 equiv. (g/MJ), as shown in Fig. 7.7. The results variation was attributed to the fact that the

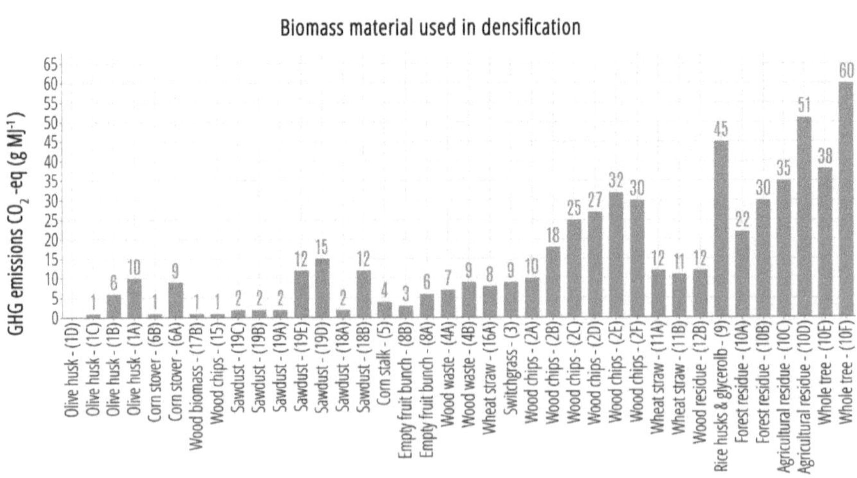

Figure 7.6 Life cycle greenhouse gas emissions in CO_2 equiv. (g/MJ) for a reference gate-to-gate biomass densification system (Reproduced with permission from Muazu et al., 2017).

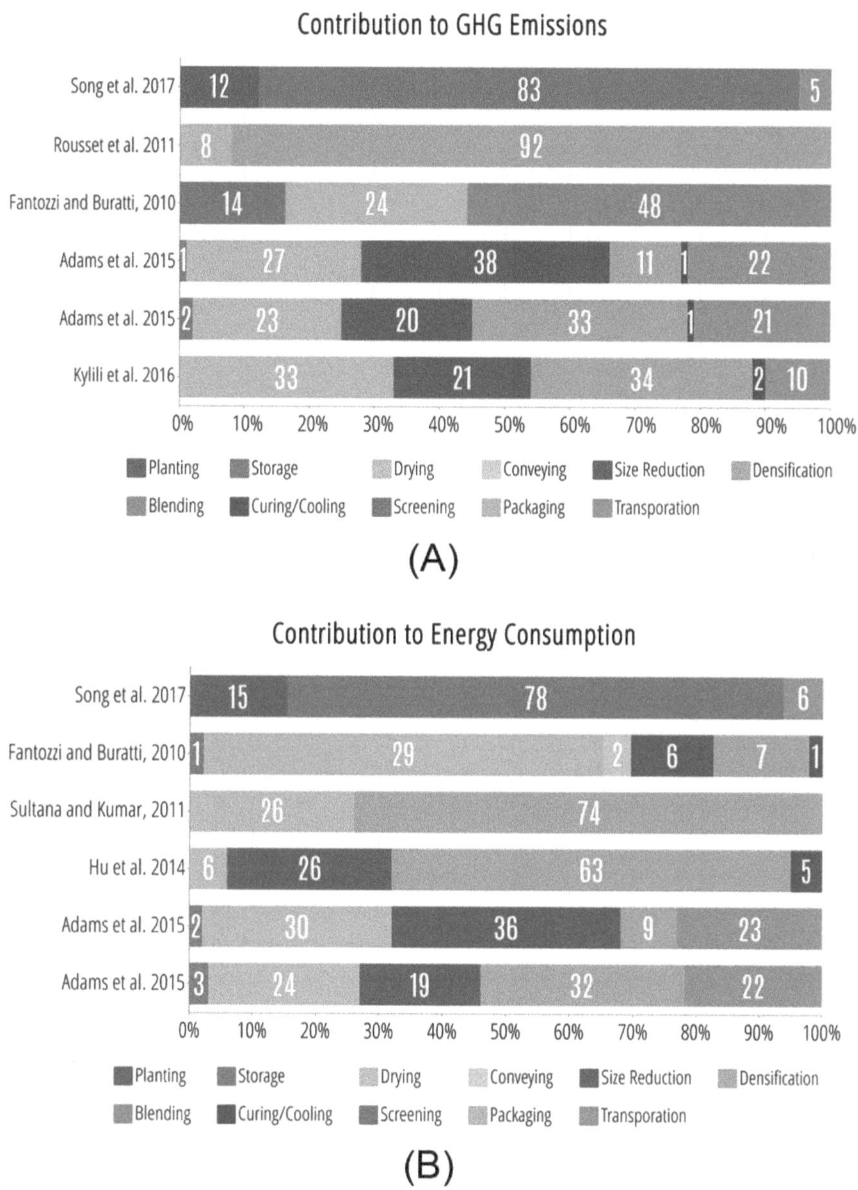

Figure 7.7 Contribution of life cycle stages to the overall greenhouse gas emissions and energy consumption of densification systems.

studies were based on data from different sources and of different quality. Some studies were based on the data from real densification plants and whether in some cases, the data had been retrieved from literature.

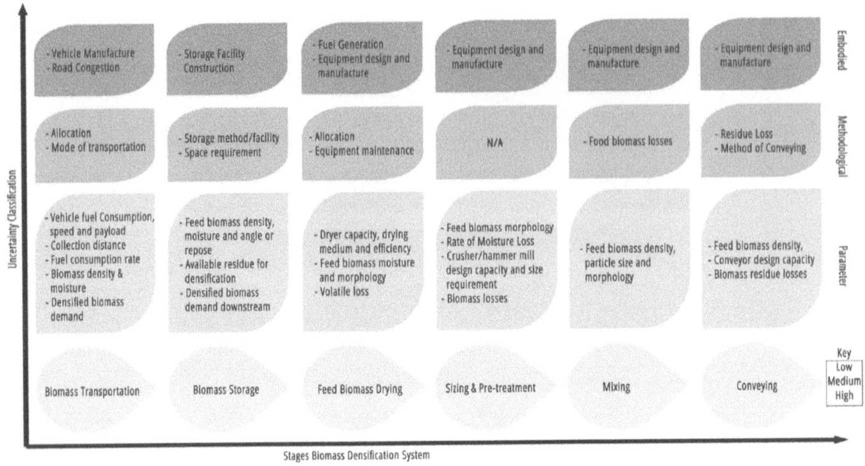

Figure 7.8 Possible sources of uncertainty in fuel densification life cycle assessment—Predensification (Reproduced with permission from Muazu et al., 2017).

The study also highlighted the parameters that affect uncertainty in biomass densification LCA. Considering a "gate-to-gate" biomass densification LCA with the transportation process being within the scope, the following classification and ranking of possible sources of uncertainty have been proposed:

- Parameter uncertainty: concerns the uncertainty that can arise due to errors in the definition of the densification phase, the relevant process inputs, the technologies used, their specific characteristics, and the relevant emissions.
- Methodological uncertainty refers to the methodology followed and the assumptions applied in order to evaluate the impact of the densification procedure, including scenario, modeling, and decision uncertainty.
- Embodied impact uncertainty is related to embodied environmental impacts of the processes involved in the densification system, such as steel production and electricity generation.

Potential sources of uncertainty during the predensification and postdensification phases are graphically presented in Figs. 7.8 and 7.9 respectively.

As a conclusion, the authors suggested the LCA studies be reported with the associated uncertainties. This would increase the clarity of the results. The expansion of analysis in similar LCA studies in order to cover a wider and more detailed range of potential environmental impacts was also recommended.

7.3.3.1 *Life cycle assessment of woody pellets/briquettes*

Laschi et al. (2016) aimed to evaluate the environmental impacts related to WP production for domestic heating using LCA on a cradle-to-gate basis;

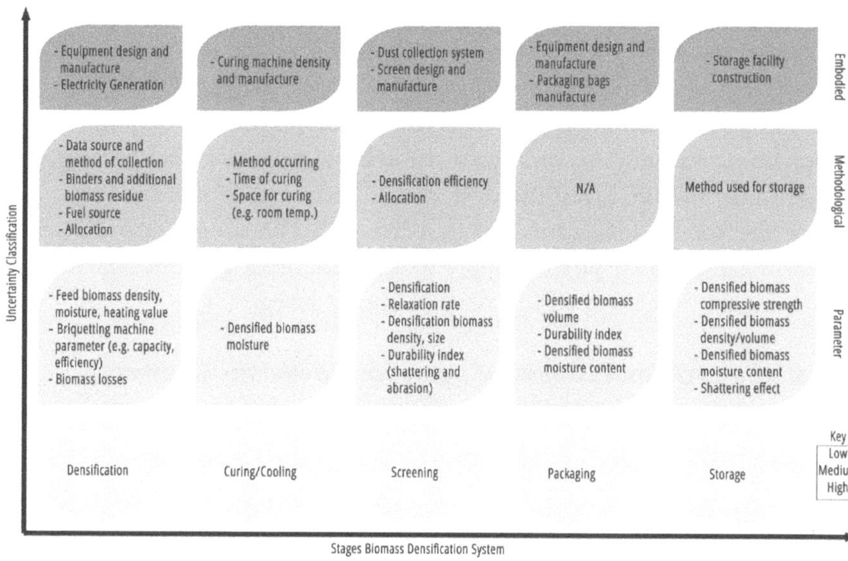

Figure 7.9 Possible sources of uncertainty in fuel densification life cycle assessment—Postdensification (Reproduced with permission from Muazu et al., 2017).

thus, considering all the processes involved, from wood extraction to packaging pellets. Different scenarios were conducted using a FU of 1 kg of packed high-quality WPs in terms of seven impact categories. The results showed the significant contribution of electricity to the pellet system, which accounted for more than 90% of the total environmental footprint. Through the alternative scenarios in the study, different means for the production of electricity and heat were proposed. It was observed that the substitution of the boiler by a cogeneration unit led to a substantial reduction of the total impact.

Ruiz et al. (2018) examined different scenarios of wood pelleting, referenced to a FU of a product, namely 1 MJ of thermal energy produced by pellets in a domestic boiler. The alternative scenarios investigated different designs of the pellet production plant, different biomass feedstocks of various origins. According to the results, the pellet making process presents the highest environmental impact, followed by the production phase of biomass, the transportation phase, and finally, the end-use phase.

Hamedani et al. (2019) studied the environmental impact of the pellet production system from vine and olive grove woody biomass from a cradle-to-gate perspective. The results of the two studies were compared using 1 MJ of pellet as a reference unit and assuming that the processed biomass feedstock is not a by-product but a coproduct. It was revealed that olive-derived pellets give a higher energy return rate than vine pellets. Also, the production phase, and specifically fertilizers and pesticides application, gave the highest contribution to all impact categories.

In their study, de la Fuente et al. (2018) aimed to quantify the environmental impact of a mobile pellet production system using logging residues in Nordic conditions. According to the results, pelleting, transporting, and drying stages in the investigated scenarios accounted for 62%, 14%, and 14% of the total GHG emissions, respectively. The use of grid electricity instead of diesel-based power supply also contributed to reducing environmental impacts in almost all the assessed impact categories.

7.3.3.2 Life cycle assessment of nonwoody pellets/briquettes

The study conducted by Kylili et al. (2016) thoroughly investigated the environmental impact of a pilot pelleting plant using olive husk, an olive oil production residue, as raw feedstock. The analysis was done on a gate-to-gate basis but also taking into consideration the transportation of raw feedstock and the densified product. Based on a novel mathematical model and using nonlinear programming, the authors developed different scenarios and, through the LCA, aimed to identify the optimal location for a set of olive husk management facilities. The impact of renewable energy sources integration with the existing pellet plant equipment on the LCA results was also examined. The analysis results confirmed the significant role that transportation plays in solid biofuels production chains and verified the extent of reduction in primary energy needs and consequently on the environmental footprint.

The environmental impact of olive pomace utilization was also studied by Duman et al. (2020) by employing a gate-to-grave approach. The investigated scenarios were based on the different olive oil production technologies, namely the traditional, three-phase, and two-phase. The scenarios were also considered as three different utilization methods: the production of pellets, fogging additives, and compost production. Among the developed scenarios, pellet production using olive pomace presented the lowest environmental burden, with traditional or two-phase production technologies being the most favorable.

In their study, Wang et al. (2017) used LCA methodology to assess the environmental impact of briquetted production using corn stalk as biomass feedstock in a cradle-to-grave system of boundaries. The LCI of the study was developed based on an actual briquette plant. According to the results, climate change and the effects of depletion of corn briquette fossils are of an order of magnitude lower than that of coal, thus 11 g CO_2 equiv. /MJ and 2 g oil equiv. /MJ, respectively.

Hu et al. (2014) conducted a comprehensive assessment of the economic, environmental, and social impact of a fully operating corn stalk briquetting plant. The authors confirmed the suitability of flat die briquetting for agricultural biomass feedstock. They noted that briquetting was found to be the most energy-consuming phase of the plant, accounting for approximately 93% of the total energy consumption. GHG emissions of the briquetting life cycle were significantly lower than coal, namely 1 kg CO_2 equiv./GJ over 157 kg CO_2 equiv./GJ, respectively. The results align with the study of Song et al. (2017), where a reduction of 90.46% of GHG emissions was reported to be potentially achieved through corn straw-pellet production and replacing coal firing.

7.3.3.3 Life cycle assessment of combined densification and torrefaction systems

A concept that receives increased interest in the last decade combines torrefaction and densification (Kabir and Kumar, 2012; Larsson et al., 2013; Uslu et al., 2008; Agar et al., 2015; Pirraglia et al., 2013).

The concept was initially proposed by Bergman (2005). Torrefaction, a mild pyrolysis pretreatment process where the raw biomass is subjected to temperatures around 220°C–300°C in a nonoxidized environment under atmospheric pressure conditions for 10 minutes to 2 hours using low heating rates, is followed by pelletization. The main product of torrefaction is a solid energy carrier called biochar, which is homogeneous and presents hydrophobic characteristics, making it more attractive for storage. Biochar maintains approximately 70% of the initial mass and 90% of the energy content of the raw feedstock; thus, it is the energy carried with improved calorific value. Through the pelletization phase that follows, an energy-dense solid biofuel is obtained, combining the benefits of both processes (Christoforou and Fokaides, 2018).

A comprehensive LCA study focused on the combined torrefaction/pelletization concept was published by Adams et al. (2015). The study aimed to compare the environmental impact of a pellet production chain with integrated torrefaction over a conventional pellet production chain. Forest-derived woody biomass feedstock was processed. A FU of 1 ton or 1 MJ of delivered pellets (torrefied or not) was employed to obtain comparative results from the two individual case studies on a cradle-to-gate basis.

The results indicated that the torrefied pellets have a lower impact on climate change on an MJ-delivered basis compared to nontorrefied products (i.e., 17.5 g CO_2 equiv. /MJ and 27.6 kg CO_2 equiv./MJ, respectively) under the low drying-energy requirement scenario. A comparison of different scenarios, where the waste torrefaction gases (i.e., torgas) were utilized as a preheating medium, contributing to the drying process of the feedstock, was also conducted. The importance of the assumptions related to the calorific value of the utilized torgas was highlighted through a sensitivity analysis. In terms of land use, it was shown that in the case of torrefied pellets, the requirements for a land area to produce the same amount of raw material (i.e., 60 kton woodchips per annum) were higher than the nontorrefied pellets scenario. Regarding the pelleting process per se, lower contribution on the impact categories of climate change and fossil fuel depletion was retrieved for torrefied pellets.

Agar et al. (2015) investigated the environmental footprint (i.e., energy input and air emissions) of torrefied pellets derived from logging residues, considering an end-use scenario where the torrefied pellets were cofired with coal. The scenario considered the production of pellets in far northern Europe and their transportation to the south. Using sustainability criteria of the renewable energy directive, final emissions related to electricity demand were around 45 g/MJ for torrefied pellets compared to 43 g/MJ for conventional pellets. Cofiring with coal would result in 77% and 78% emissions savings for torrefied and nontorrefied pellets, respectively, compared to the emissions from fossil coal. Although torrefied pellets presented slightly higher emissions, the physical characteristics of the end-product enable a higher percentage of cofiring with coal, thus enhancing higher energy savings.

Alanya-Rosenbaum and Bergman (2019) carried out a cradle-to-grave LCA of nontorrefied briquettes and torrefied briquettes (TOB) derived from postharvest forest residues. The study aimed to compare the impact of biomass conversion near the woods in comparison with an in-town facility and its utilization for electricity or heat generation. Thus, two FUs were used, namely 1 MJ of useful heat produced for domestic heating and 1 kWh of electricity generated at a power plant. Within the 48%–55% range, lower global warming impact was obtained for biomass processing near the source for both end-uses. Also, in the case of TOB, the torrefaction process was found as the main contributor to the global warming impact category.

7.4 Conclusions

LCA studies of solid biofuel conversion systems, including biomass densification, are more than necessary to identify the hotspots with regard to their environmental impact. The outcome of those studies consists a key element to the promotion of more sustainable production/conversion systems, contributing at the same time to the achievement of the global bioenergy targets.

Although a standardized methodology guides the framework of LCA studies, the variations in the definition of system boundaries, the uncertainty of data in many cases, as well as the assumptions applied always make the comparison of different studies quite challenging and sometimes irrelevant. Therefore, each LCA's goal, scope, and system boundaries should be clear and specific, and the relevant inputs and outputs in the system well defined. The results of previous LCA studies in the field of biomass densification highlighted the significant impact the pelleting/briquetting stage has in such systems, as well as the contribution of transportation to their environmental footprint. Therefore, a more detailed investigation of biomass densification systems shall be promoted and encouraged in the future.

References

Adams, P.W.R., Shirley, J.E.J., McManus, M.C., 2015. Comparative cradle-to-gate life cycle assessment of wood pellet production with torrefaction. Applied Energy 138, 367–380.

Agar, D., Gil, J., Sanchez, D., Echeverria, I., Wihersaari, M., 2015. Torrefied vs conventional pellet production—a comparative study on energy and emission balance based on pilot-plant data and EU sustainability criteria. Applied Energy 138, 621–630.

Alanya-Rosenbaum, S., Bergman, R.D., 2019. Life-cycle impact and exergy based resource use assessment of torrefied and non-torrefied briquette use for heat and electricity generation. Journal of Cleaner Production 233, 918–931.

Bergman, P.C.A., 2005. Combined Torrefaction and Pelletisation - the Top Process. ECN report, ECN-C-073.

Bioenergy Europe, 2019, Statistical Report-Pellets.

Chen, L., Xing, L., Han, L., 2009. Renewable energy from agro-residues in China: solid biofuels and biomass briquetting technology. Renewable and Sustainable Energy Reviews 13 (9), 2689–2695.

Christoforou, E., Fokaides, P.A., 2018. Advances in Solid Biofuels. Springer.

de la Fuente, T., Bergström, D., González-García, S., Larsson, S.H., 2018. Life cycle assessment of decentralized mobile production systems for pelletizing logging residues under Nordic conditions. Journal of Cleaner Production 201, 830–841.

Duman, A.K., Özgen, G.Ö., Üçtuğ, F.G., 2020. Environmental life cycle assessment of olive pomace utilization in Turkey. Sustainable Production and Consumption 22, 126–137.

Ecoinvent Centre, 2017. Ecoinvent data v3.0. <http://www.ecoinvent.org> (accessed 26 March 2020).

El Hanandeh, A., 2015. Energy recovery alternatives for the sustainable management of olive oil industry waste in Australia: life cycle assessment. Journal of Cleaner Production 91, 78–88.

European Life Cycle Database (ELCD), 2006. Joint Research Center. <http://eplca.jrc.ec.europa.eu/ELCD3/> (accessed 26 March 2020).

Fantozzi, F., Buratti, C., 2010. Life cycle assessment of biomass chains: wood pellet from short rotation coppice using data measured on a real plant. Biomass and Bioenergy 34 (12), 1796–1804.

Fokaides, P.A., Christoforou, E., 2016. Life cycle sustainability assessment of biofuels. Handbook of Biofuels Production. Woodhead Publishing, pp. 41–60.

Hamedani, S.R., Colantoni, A., Gallucci, F., Salerno, M., Silvestri, C., Villarini, M., 2019. Comparative energy and environmental analysis of agro-pellet production from orchard woody biomass. Biomass and Bioenergy 129, 105334.

Hu, J., Lei, T., Wang, Z., Yan, X., Shi, X., Li, Z., et al., 2014. Economic, environmental and social assessment of briquette fuel from agricultural residues in China—a study on flat die briquetting using corn stalk. Energy 64, 557–566.

Intelligent Energy Europe (IEE), 2011. BioGrace - harmonized calculations of biofuel greenhouse gas emissions in Europe. <www.biograce.net> (accessed 26 March 2020).

International Organization for Standardization (ISO). ISO 14044:2006. Environmental management - Life cycle assessment - requirements and guidelines, 2006.

International Organization for Standardization (ISO) 2006, ISO 14040:2006: Environmental management - Life cycle assessment - Principles and framework, 2006.

Kabir, M.R., Kumar, A., 2012. Comparison of the energy and environmental performances of nine biomass/coal co-firing pathways. Bioresource Technology 124, 394–405.

Karkania, V., Fanara, E., Zabaniotou, A., 2012. Review of sustainable biomass pellets production - A study for agricultural residues pellets' market in Greece. Renewable and Sustainable Energy Reviews 16 (3), 1426–1436.

Kirsten, C., Lenz, V., Schröder, H.W., Repke, J.U., 2016. Hay pellets—the influence of particle size reduction on their physical–mechanical quality and energy demand during production. Fuel Processing Technology 148, 163–174.

Klemm, M., Schmersahl, R., Kirsten, C., Weller, N., 2013. Biofuels biofuel: upgraded new solids biofuel upgraded new solids. Renewable Energy Systems. Springer, pp. 138–160.

Kylili, A., Christoforou, E., Fokaides, P.A., 2016. Environmental evaluation of biomass pelleting using life cycle assessment. Biomass and Bioenergy 84, 107–117.

Larsson, S.H., Rudolfsson, M., Nordwaeger, M., Olofsson, I., Samuelsson, R., 2013. Effects of moisture content, torrefaction temperature, and die temperature in pilot scale pelletizing of torrefied Norway spruce. Applied energy 102, 827–832.

Laschi, A., Marchi, E., González-García, S., 2016. Environmental performance of wood pellets' production through life cycle analysis. Energy 103, 469–480.

Muazu, R.I., Borrion, A.L., Stegemann, J.A., 2017. Life cycle assessment of biomass densification systems. Biomass and Bioenergy 107, 384–397.

Pirraglia, A., Gonzalez, R., Saloni, D., Denig, J., 2013. Technical and economic assessment for the production of torrefied ligno-cellulosic biomass pellets in the US. Energy Conversion and Management 66, 153–164.

Quinteiro, P., Tarelho, L., Marques, P., Martín-Gamboa, M., Freire, F., Arroja, L., et al., 2019. Life cycle assessment of wood pellets and wood split logs for residential heating. Science of the Total Environment 689, 580–589.

Ruiz, D., San Miguel, G., Corona, B., López, F.R., 2018. LCA of a multifunctional bioenergy chain based on pellet production. Fuel 215, 601–611.

Shahrukh, H., Oyedun, A.O., Kumar, A., Ghiasi, B., Kumar, L., Sokhansanj, S., 2015. Net energy ratio for the production of steam pretreated biomass-based pellets. Biomass and Bioenergy 80, 286–297.

Song, S., Liu, P., Xu, J., Chong, C., Huang, X., Ma, L., et al., 2017. Life cycle assessment and economic evaluation of pellet fuel from corn straw in China: a case study in Jilin Province. Energy 130, 373–381.

US Life Cycle Inventory Database, 2012. National Renewable Energy Laboratory. <https://www.lcacommons.gov/nrel/search> (accessed 26 March 2020).

Uslu, A., Faaij, A.P., Bergman, P.C., 2008. Pre-treatment technologies, and their effect on international bioenergy supply chain logistics. Techno-economic evaluation of torrefaction, fast pyrolysis and pelletisation. Energy 33 (8), 1206–1223.

Wang, Z., Lei, T., Yang, M., Li, Z., Qi, T., Xin, X., et al., 2017. Life cycle environmental impacts of cornstalk briquette fuel in China. Applied Energy 192, 83–94.

CHAPTER 8

Life cycle assessment of geothermal power technologies

Andrea Paulillo[1], Alberto Striolo[2] and Paola Lettieri[1]
[1]Department of Chemical Engineering, University College London, London, United Kingdom
[2]School of Chemical, Biological and Materials Engineering, University of Oklahoma, Norman, OK, United States

Contents

8.1 Introduction

The word geothermal derives from the Greek *geo*, meaning earth, and *thermos*, meaning hot. Geothermal energy is the natural heat of the Earth that originated from its formation and is continuously replenished by the decay of radioactive substances, primarily uranium, thorium, and potassium. Geothermal heat flows from the Earth's interior to the surface, eventually escaping to outer space by means of convection and

Environmental Assessment of Renewable Energy Conversion Technologies
DOI: https://doi.org/10.1016/B978-0-12-817111-0.00002-4

conduction at a rate of approximately 44 TW (Gando et al., 2011). Although this quantity is more than double humanity's current average power consumption from primary sources, most of it is too diffuse to be economically recoverable. The average geothermal heat flux at the Earth surface is ~0.09 W/m^2; by comparison, solar radiation reaches the Earth at an average rate of ~160 W/m^2 (Wild et al., 2013). Unlike geothermal energy, solar energy is intermittent and dependent on climatic conditions.

Although the total heat content of the Earth is not unlimited, geothermal energy is regarded as renewable because it is tremendously higher than current human consumption—at current consumption rates, it would take humankind around 10^9 years to exhaust the Earth's geothermal resources (Rybach, 2007). However, local depletion of geothermal energy can occur when the extraction rate is greater than the ability of the resource to recover (Rybach, 2003, 2007). For this reason, the lifetime of a geothermal power plant is typically estimated to be 30 years.

Geothermal energy in the form of hot springs has been used since Paleolithic times for bathing; historically, geothermal energy has also been used to heat buildings in many locations, from China to Rome (Cataldi et al., 1999). Direct heat remains the most dominant form of geothermal use, ranging from district heating, industrial use, agriculture, and heat pumps. The use of geothermal energy for the generation of electricity started at the beginning of the 20th century: the first demonstration plant was built in Larderello of Tuscany, Italy, in 1904, which was followed by a commercial power generation plant in 1911 (Parri and Lazzeri, 2016). The contribution of geothermal energy to global electricity production is still minute: in 2018, geothermal energy generated approximately 90 t Wh, ~0.3% of global electricity generation (IEA, 2019a, 2019b). Today, the largest producers are the United States, the Philippines, and Indonesia (Bertani, 2016).

The simplest way to classify geothermal resources is according to their enthalpy. Medium to high enthalpy reservoirs, typically with temperatures above 100°C−150°C, are used for electricity generation, while low enthalpy reservoirs are devoted to direct uses. A detailed classification of geothermal reservoirs is provided by Lund (2007), adapted from White and Wiliams (1975). Traditional geothermal resources (i.e., convective hydrothermal systems) for power generation are located at geological plate boundaries, which are loci of volcanic activity and thus offer high heat flows. In these locations, geothermal heat moves upward by the convective circulation of naturally occurring hot water or steam through rock formations of sufficient permeability. Unconventional geothermal resources explored for power generation are found in regions with higher than average geothermal gradient where water and/or sufficient rock permeability is absent (i.e., dry/wet hot rock resources) and in regions with near-surface granitic intrusions that heat up local groundwater through the decay of radioactive elements (radiogenic resources).

Geothermal energy is expected to expand substantially in the coming years along with other renewable energy sources, mainly to mitigate climate change by reducing carbon emissions. In the four-year period from 2015 to 2018, power generation via geothermal energy grew by ~15% (from 81 terawatt-hours), increasing by ~3 t Wh per year; it is projected to grow further 28% by 2024, to reach 116 t Wh. However, such projected expansion falls short of that envisaged by the International Energy Agency (IEA) within their "Sustainable Development Scenario," which would require a production level of 162 t Wh by 2025, that is, an increase of nearly 75% with respect to current levels (IEA, 2019a, 2019b).

The key advantage of geothermal energy over other renewable energy sources is that it is independent of seasonal and climatic variations, representing a base-load electricity source. Geothermal energy is also relatively cheap, with a total cost per unit of electricity (~0.07 USD/kWh) comparable with fossil fuels, onshore wind, hydropower, and lower than solar photovoltaic and offshore wind (IRENA, 2019). With about a quarter of the cost due to the initial investments connected with site exploration and wells drilling, securing capital funding is the main impediment to further geothermal developments.

Given that every anthropogenic activity has an environmental footprint, to identify environmentally conscious energy portfolios, it is imperative to quantify the environmental footprint of power generation via geothermal operations over their lifetimes and compare it against that of other energy sources. Life cycle assessment (LCA) is the most widely used methodology for assessing the environmental impacts of products or services over their entire life cycle. LCA is only concerned with potential impacts arising from routine operations; impacts arising from stochastic events such as accidents are outside the domain of LCA and are typically addressed by risk assessment. The application of LCA to the geothermal industry can identify the most environmentally efficient technologies, pinpoint key areas for improvements, and support the development of novel technologies.

The LCA standardized framework (ISO, 2006a, 2006b) is not tailored to specific technologies; instead, it is generally applicable. This leaves LCA practitioners with an extensive range of methodological choices that can lead to substantial differences in the estimates of the environmental performance, especially for least known and emerging technologies. This chapter is intended to provide methodological indications for the application of LCA to geothermal power systems. The chapter covers the most sensitive aspects of each phase of the standardized LCA framework (ISO, 2006a, 2006b). Section 8.2 presents a brief overview of the technologies for power generation from geothermal resources, while Section 8.3 deals with the methodological aspects of LCA. In Section 8.4, a few case studies are summarized.

8.2 Technologies for power generation

The selection of the most appropriate technology for power generation depends on the properties of the geothermal resource, primarily its enthalpy and physical state. The use of geothermal resources has historically proceeded from the highest quality and least common resources to the lowest quality and most common resources, essentially from dry-steam reservoirs to liquid-dominated reservoirs to hot rock and radiogenic resources. In this order, the following subsections present a brief overview of geothermal power technologies (see Table 8.1), including mature technologies like dry-steam, single and multistage flash, and binary cycle plants, as well as enhanced geothermal systems (EGS), which have been receiving increasing attention in recent years. A detailed description of each power generation technology is given by DiPippo (2016a, 2016b).

8.2.1 Dry-steam technology

Dry-steam geothermal reservoirs are of the highest quality and also the rarest geothermal reservoirs. It was estimated that only about 5% of all conventional hydrothermal systems with temperatures above 200°C are dry-steam type (White, 1973). Larderello in Tuscany, Italy, and The Geysers in California, US, are the only two large reservoirs of this type that have been discovered (DiPippo, 2016c). Although in 2015 dry-steam

Table 8.1 Technologies for power generation covered in this section.

Conventional resources	Dry-steam Single-flash Multistage flash Binary cycle
Unconventional resources	Enhanced geothermal systems

plants accounted for 10% of the total number of geothermal plants worldwide, they represented 23% of the total installed capacity of geothermal plants because they are characterized by an above-average capacity per generation unit of 42 MW (Bertani, 2016; IGA, 2015). Because the geothermal fluid consists solely of steam, dry-steam plants are the simplest and the cheapest technology for geothermal power generation. That is why they were the first type of geothermal technology to achieve commercial status at Larderello in 1911 (Parri and Lazzeri, 2016).

Fig. 8.1 illustrates a schematic diagram of a generic dry-steam power plant. The steam gathered from the wells is first treated to remove entrained particulate matter and small liquid droplets. The power cycle consists of a steam turbine, a generator for electricity production, a condenser (direct-contact or surface-type), and an associated cooling tower. The steam expands in the turbine converting thermal energy into mechanical energy, which is then transformed into electricity in the generator. The exhaust steam from the turbine is condensed; the condenser lowers the pressure at the outlet of the turbine, thus increasing the overall

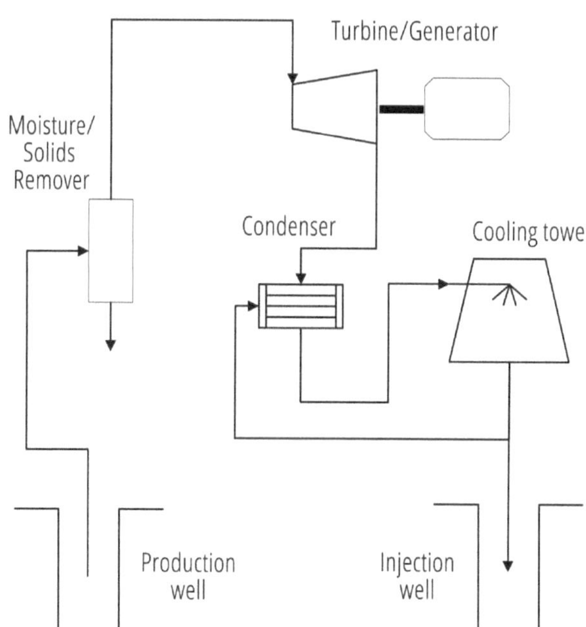

Figure 8.1 Simplified flow diagram of dry-steam power plant. Source: *Adapted from DiPippo, R., 2016a. Geothermal Power Plants: Principles, Applications, Case Studies and Environmental Impact, 4th ed. Butterworth-Heinemann.*

energy conversion efficiency. The cooling tower uses makeup water from the condensed stream to transfer waste heat to the atmosphere while providing cooling water to the condenser. Finally, the remaining condensed stream is typically reinjected into the geothermal reservoir.

Depending on the chemical composition of the steam—typically around 90% water and 10% noncondensable gases (NCGs) primarily including CO_2, H_2S, and NH_3 (Santoyo et al., 1991)—the power plant represented in Fig. 8.1 may require a gas extraction system such as vacuum pumps and stream-jet ejector to remove NCGs (Phair, 2016). This step is required for three reasons. First, NCGs can accumulate in the condenser, thus reducing its efficiency. Second, high concentrations of hydrogen sulfide (H_2S) can lead to corrosive damages to piping and turbine components. Third, discharges of NCGs into the atmosphere can cause significant environmental impacts. CO_2 and sometimes methane (CH_4) are the most significant contributors to the carbon footprint of dry-steam and single and multistage flash plants (e.g. see Paulillo et al., 2019b; and see Section 8.4). H_2S is a noxious gas that can lead to respiratory problems and acidification of freshwater systems. Many countries (e.g., Italy), having set regulatory limits on the amount of H_2S that can be released, require other abatement systems like the AMIS system implemented at Larderello (Baldacci et al., 2005).

8.2.2 Single-flash technology

In most cases, the geothermal fluid is extracted as a mixture of liquid and steam. The power plant requires an additional step to separate the steam from the liquid phase in such circumstances. In 2015, single-flash plants, including this additional step, accounted for 41% of worldwide geothermal power capacity, with an average unit capacity of 28 MW (Bertani, 2016; IGA, 2015).

A simple diagram of a generic single-flash plant is provided in Fig. 8.2. The fluid mixture of gas and liquid is separated in a high-pressure separator, typically a cylindrical cyclonic pressure vessel, where the phases are efficiently separated due to a difference in densities (DiPippo, 2016d). The steam undergoes the same steps as in dry-steam power plants; it first expands in a turbine and then condenses before reinjection. Reinjection of the brine is fundamental for two reasons: first, because it contains high concentrations of dissolved minerals that existed in the original geothermal fluid (including lithium, magnesium, potassium, arsenic, and others), which can be detrimental to surface and groundwaters; second, because it

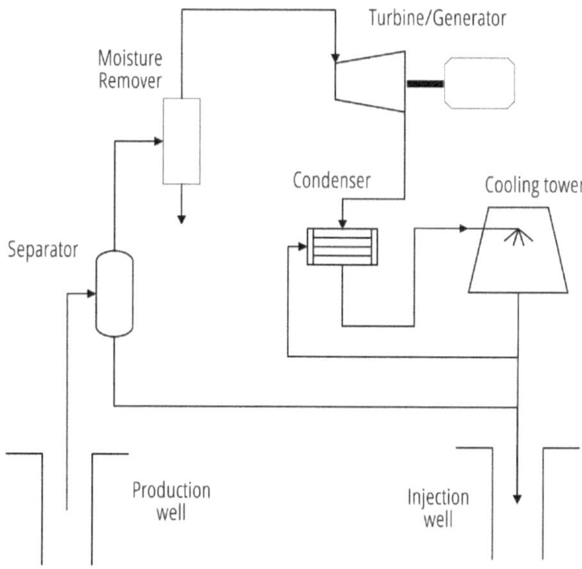

Figure 8.2 Simplified flow diagram of single-flash power plant. Source: *Adapted from DiPippo, R., 2016b. Geothermal Power Generation: Developments and Innovation. Woodhead Publishing.*

restores some of the fluid that was extracted, thus providing pressure support to the geothermal reservoir.

The production of a brine stream from the separator provides commercial opportunities. The thermal energy of the brine can be exploited in various manners, for example, via district heating (as it is extensively done in Iceland) or in greenhouses (Dunstall and Foster, 1998). Geothermal plants that cogenerate heat and power pose methodological challenges in LCA when the objective of the study is to quantify the environmental impacts of either electricity or heat production. The brine could also represent a valuable source of important metals like lithium and rare-earth elements (Smith et al., 2017). For instance, the feasibility of lithium extraction is currently being explored in Cornwall, UK, by Cornish Lithium and in Soultz-sous-Forêts, France, by EuGeLi (European Geothermal Lithium brines).

Furthermore, single-flash plants provide opportunities for mixed and combined cycles whereby the main power cycle is combined with an additional binary cycle to harness the remaining heat either in the separated liquid phase or the exhaust condensed phase. Single-flash plants share the same challenges associated with managing NCGs discussed in relation to dry-steam plants.

8.2.3 Multistage flash technologies

Double and triple-flash plants are the logical extension of single-flash plants. The additional flash stages increase the power output per unit of geothermal fluid, which often justifies the increased installation and maintenance costs. Double-flash plants have an average capacity of approximately 25% higher than that of single-flash plants (35 MW), while triple-flash plants reach 72 MW. In 2015, double-flash plants accounted for 19% of worldwide geothermal capacity, while triple-flash plants for only 1% (Bertani, 2016; IGA, 2015).

As shown in Fig. 8.3, the key difference between single- and multistage flash plants is represented by the expansion stage, which needs to accommodate multiple vapor streams produced by the additional flash stages at decreasing pressure. The industrial practice envisages using either a single multiple-admission turbine, accommodating streams at different, or multiple turbines working at different pressures (DiPippo, 2016e).

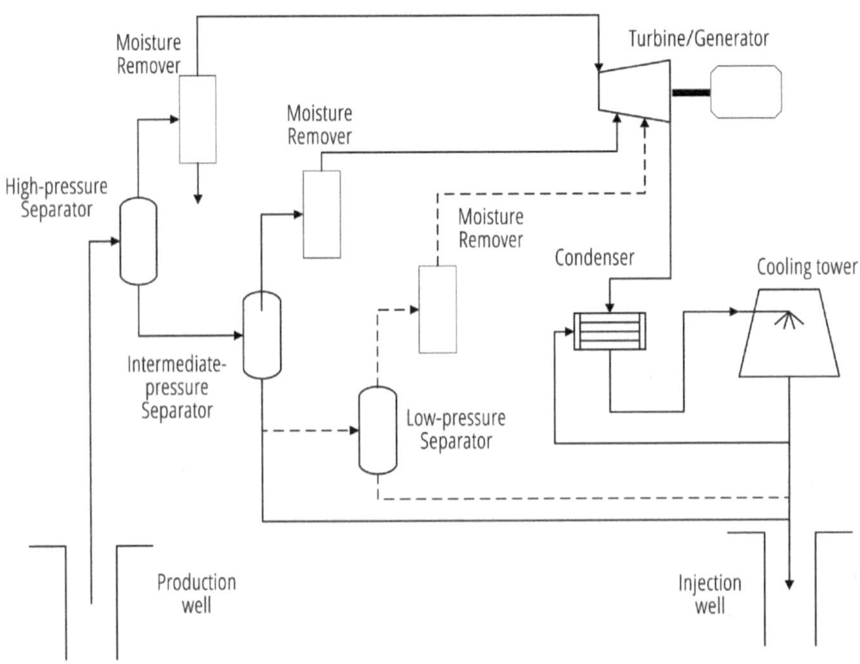

Figure 8.3 Simplified flow diagram of double- and triple-flash power plants. The dotted line refers to the triple-flash configuration. Source: *Adapted from DiPippo, R., 2016c. Dry-steam power plants. In: DiPippo, R. (ed.), Geothermal Power Plants: Principles, Applications, Case Studies and Environmental Impact. Butterworth-Heinemann.*

Multistage operations can be implemented in mixed and combined cycles like single-flash plants and can cogenerate electricity and thermal energy. In multistage flash plants, the brine can be very highly concentrated and thus present a higher potential for mineral recovery. However, it also has a higher potential for yielding scaling problems (Harvey and Wallace, 2016).

8.2.4 Binary cycle technology

Binary cycles are used for liquid-dominated reservoirs with temperatures lower than 200°C (more often 150°C) where single and multistage flash plants are less efficient and consequently less economical. The average capacity of these plants is about 6 MW per unit. This, lower than average capacity, explains why although accounting for 46% of the total number of geothermal power units, they only yield 14% of the total installed capacity (Bertani, 2016; IGA, 2015). However, units with higher ratings of ~20 MW are being developed through advanced binary cycle designs like dual-pressure, dual-fluid, and Kalina binary cycles (DiPippo, 2016f).

Fig. 8.4 illustrates a simplified diagram of a generic binary cycle plant. At low temperatures, the geothermal fluid is unlikely to flow spontaneously. The potential of scaling is considerable due to the low temperature. To promote flow, binary cycle plants use downhole pumps to produce pressurized fluid. The thermal energy of the pressurized geothermal fluid is transferred to a secondary working fluid with a low boiling point (e.g. propane, butane, and pentane), which performs a closed power cycle similar to those used in conventional fossil-fuel-powered and nuclear plants. The heated working fluid expands in a turbine to generate electricity; it is then condensed and repressurized to close the cycle.

Once it has heated the working fluid, the geothermal fluid is reinjected into the geothermal reservoir. For this reason, geothermal binary plants are typically the most environmentally benign: they have no discharges of NCGs. The greatest environmental threat is associated with the loss of the working fluid, which may be a potent greenhouse gas, as in the case of refrigerants such as R32, R134. Note that potent ozone-depleting substances such as R-12 and R-114 have been banned from use by the Copenhagen Amendment to the Montreal Protocol.

Although binary cycle plants can extract more energy from the geothermal fluid than the other technologies described above, they are only used at moderate temperatures. This happens because current downhole pumps are limited in operating temperature and setting depth and prevent excessive minerals precipitation in heat exchangers and the wells (Mines, 2016).

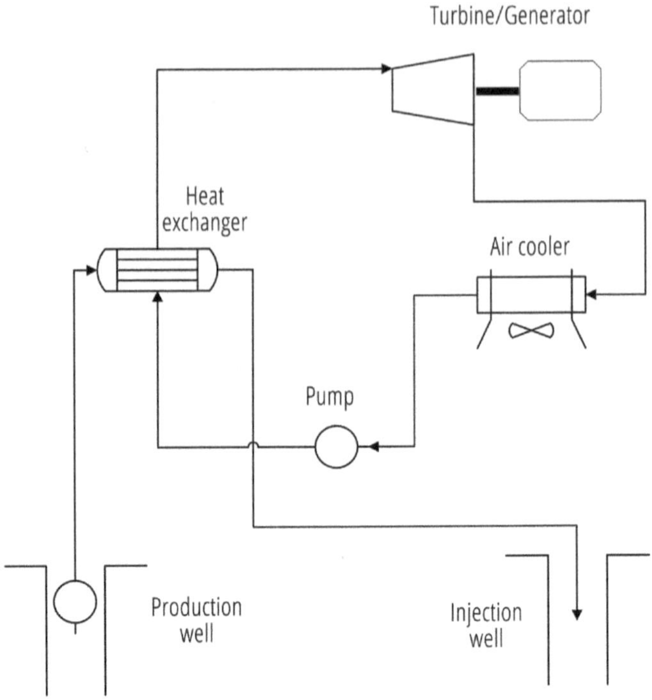

Figure 8.4 Simplified flow diagram of binary cycle power plant. Source: *Adapted from DiPippo, R., 2016f. Binary cycle power plants. In: DiPippo, R. (ed.), Geothermal Power Plants: Principles, Applications, Case Studies and Environmental Impact. Butterworth-Heinemann.*

8.2.5 Enhanced geothermal systems

The EGS concept originates in the early 1970s from the "Hot Dry Rock" project at the Fenton Hill site in New Mexico (US) (Breede et al., 2013). The first commercial EGS plants were deployed in Europe at the beginning of the 21st century, at Soultz-sous-Forêts in France and Landau in Germany (Lu, 2018). Today, there are active EGS projects in many countries, including the UK, Australia, and Japan. The contribution of EGS to total geothermal capacity is still negligible, with an installed capacity of commercial EGS plants in the range of 1−10 MW.

Over the years, different definitions of EGS have been proposed, covering a wide variety of rock types, depth, temperature, reservoir permeability and porosity, and type of stimulation technique involved. This chapter adopts the definition of the MIT, which defines EGS as "engineered reservoirs that have been created to extract economic amounts of heat from low permeability and/or porosity geothermal resources," including "all geothermal resources that are currently not in commercial production and require stimulation or enhancement" (MIT, 2006).

EGS differs from conventional geothermal exploitation in that they need the creation of engineered reservoirs employing hydraulic stimulation. Water is circulated through the reservoir and used to generate power using one of the technologies described above (typically binary cycle plants) depending on the temperatures that can be harnessed (MIT, 2006). The EGS technology is also being used to expand production from existing geothermal fields such as Larderello in Italy, Coso, and The Geysers in the US (Breede et al., 2013).

8.3 Life cycle assessment: methodological aspects

This section presents a compendium of methodological aspects of LCA, as they apply specifically to the geothermal power sector. The section follows the four phases of the LCA standardized framework (ISO, 2006a, 2006b); a detailed description of each phase in a generic LCA approach is provided by Baumann and Tillman (2004) and Hauschild et al. (2018).

8.3.1 Goal and scope definition

8.3.1.1 Goal

The appropriate definition of the goal of an LCA study is paramount in that it affects all subsequent phases, in particular identification of the system boundaries and appropriate elaboration of the life cycle inventory (LCI). An LCA study's goal primarily includes the intended application of the results (i.e., *what* the study does) and the reasons for carrying out the study (i.e., *why* the study is undertaken). Both aspects affect the perspective adopted, that is, retrospective or prospective, and the LCA "mode," essentially attributional or consequential (Finnveden et al., 2009).

The majority of LCA studies on geothermal power generation adopt an attributional approach with either retrospective or prospective perspective, depending on whether the focus is on an existing or a future geothermal power plant. The consequential mode can be used, for instance, to evaluate the environmental consequences of increasing electricity production from geothermal power within a local, regional, or national community, when such an increase is significant enough to cause the so-called structural changes (see JRC, 2010). Examples of this approach still lack in the literature.

8.3.1.2 Functional unit

In LCA, the Functional Unit (FU) defines qualitative and quantitative aspects of the function or functions delivered by the product system. It also determines the reference flow, that is, the product flow to which all input and output flow for the processes in the product system must be quantitatively related. The FU is an essential concept in LCA because it enables fair and consistent comparison between different products.

The main function of a geothermal power plant is the generation of electricity at high voltage. Thus, the FU is defined as an amount of electricity expressed in kilowatt-hour or megajoule (typically 1 kWh). When the objective of the study is to assess the environmental impact of a combined-heat-and-power geothermal plant, the FU corresponds to the amount of electricity and heat generated within the same time frame. For instance, an appropriate FU for the Hellisheiði geothermal power plant corresponds to the amount of electricity and thermal energy generated within one second, which equals 303 and 133 MJ, respectively (Paulillo et al., 2019b).

8.3.1.3 System boundaries

The system boundaries demarcate the confines of the studied product system with respect to the surrounding economy and the environment. Ideally, they should be represented in a schematic diagram that illustrates the processes included in the product system. This section follows the approach developed for Environmental Product Declarations, which divides the product system into three modules: upstream, core, and downstream (Environdec, 2019).

The upstream module includes the processes for the production of materials and energy used by the core module. The core module comprises the construction of a geothermal power plant's infrastructure, operation, maintenance, and end-of-life phases. The downstream module consists of electricity distribution (and thermal energy in the case of a cogeneration plant) from the power plant to the customer. This phase is typically not included in LCA studies because (1) it has negligible impacts and (2) it does not differ between alternative methods of electricity generation.

8.3.2 Life cycle inventory: key aspects and parameters

LCI analysis is arguably the most challenging and time-consuming phase of LCA. This section presents an overview of the main aspects and parameters that are necessary for modeling the life cycle of a geothermal power plant.

8.3.2.1 Geothermal resource

The type, quality, and location of the geothermal resource affect several aspects of a geothermal power plant's construction, operation, maintenance, and end-of-life. As noted in section 8.0, the selection of the appropriate power generation technology depends on the type and quality of the geothermal reservoir. The former essentially differentiates between traditional convective hydrothermal systems and unconventional resources like dry/wet hot rocks. The latter includes the dominant phase of the reservoir and determines the temperature at which geothermal fluid is produced.

The composition of the steam, particularly of NCGs, needs to be determined or estimated when steam is the dominant phase in the geothermal fluid. The presence of NCGs such as CO_2 and H_2S affects (1) the magnitude of environmental impacts due to operational emissions from condensing the geothermal fluid and (2) the need for additional abatement systems. For binary cycles, it is also important to determine the composition of the geothermal fluid, which dictates the minimum reinjection temperature. On the other hand, the temperature of the geothermal fluid influences the efficiency of the power cycle (Zarrouk and Moon, 2014) and determines the number of production wells required to sustain the installed capacity of the power plant.

Determination of the reservoir depth is essential as it affects the depth of the geothermal boreholes. Typical depths for conventional hydrothermal reservoirs range between 2 and 4 km. EGSs wells are generally deeper, extending up to 6 km. Finally, because the construction phase and, in particular, the drilling of geothermal wells represent one of the largest sources of environmental impacts, the success rate of geothermal wells may be taken into account in estimating the environmental footprint of a new operation. Global median values for success rates of exploratory, primary, and make up wells are approximately 67%, 72%, and 79% (IFC, 2013).

8.3.2.2 Power generation technology

Section 8.0 presents the most common technologies for geothermal power generation. Irrespectively of the technology implemented, three parameters are of utmost importance for LCA studies, and these are

- the installed capacity of the plant, that is, the maximum gross power that the turbine can generate.
- the auxiliary power requirements, that is, the electric requirements of auxiliary services of the power cycle like those of pumps and cooling towers.

- the capacity (or load) factor, that is, the ratio of the actual to the maximum electricity output within a period of time.

It is also necessary to determine whether the geothermal wells are self-flowing or require pumping for liquid-dominated reservoirs. The power requirement of a downhole pump, which depends on the geothermal fluid flow rate, can significantly lower the net electricity output of the geothermal plant (and thus its environmental performance).

8.3.3 Life cycle inventory: core module phases and activities

This section describes the main activities in the core module of the life cycle of a geothermal power plant, which includes construction, operation, maintenance, and end-of-life (see Fig. 8.5).

8.3.3.1 Construction phase

- Geothermal wells

  The selection of the drilling technique depends on the temperature, host rock, gas concentration in the fluid, and depth of the geothermal reservoir. Rotary drilling is the most commonly applied technique: the drilling bit, which is attached to a string of drill pipe hung from a derrick, is rotated by an engine to apply concentrated loads on the rock, which causes it to crack and spall. A diesel-electric generating set typically powers the engine. Karlsdóttir et al. (2015) report diesel consumption for the Hellisheiði plant of ~1.5–3 MJ/m; values for EGS operations are higher, ranging from 4 and up to 7–8 MJ/m (Frick et al., 2010; Paulillo et al., 2020a, 2020b). Previous LCA studies demonstrated that indirect (from upstream) and direct (from burning) emissions from the diesel generator represent a substantial portion of the environmental impacts of the construction phase (Lacirignola and Blanc, 2013; Paulillo et al., 2020a, 2019b; Pratiwi et al., 2018). Therefore, it is suggested that (1) using electricity from the grid to power drilling rigs and (2) employing drilling

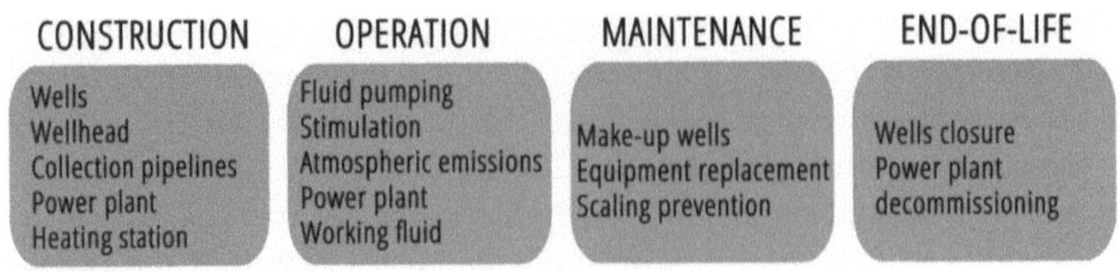

Figure 8.5 Phases and activities in the Core module.

techniques with lower specific energy consumption (like noncontact drilling technologies) would significantly improve the environmental performance of the construction phase (Menberg et al., 2016).

The drilling process requires drilling muds to remove the drilled cuttings and bring them to the surface, as well as for cooling the drill bit, lubricating the drill string, and preventing the collapse of the well during drilling. Drilling muds are made up of a mixture of water and clay (bentonite) and may also contain additives such as barium sulfate (barite), synthetic polymers, and anionic polyelectrolytes. Most of these chemicals are not included in commercial databases, so they can be modeled as generic organic chemicals (Paulillo et al., 2020a, 2019b) until the database is updated.

Geothermal wells are cased in stainless steel strings and liners throughout most of their length, and the space between casing and rock is filled with cement. The objective of the casing is to provide structural integrity to the wells and prevent unintentional losses of fluid. Indirect emissions associated with steel production have a substantial contribution to the impact of the construction phase. Typically around 80–120 kg of stainless steel and 30–100 kg of cement per meter are required (Frick et al., 2010; Karlsdóttir et al., 2015; Lacirignola and Blanc, 2013; Paulillo et al., 2020a; Pratiwi et al., 2018).

Exploratory wells, typically three (DiPippo, 2016a), are drilled at the outset of a new geothermal project to confirm the geothermal field's existence and quality. Exploratory wells have a smaller diameter than primary wells (i.e., production and injection). Marchand et al. (2015) assume a scaling factor of 0.5 and 0.3 for estimating the requirements of cement and steel, respectively, from those of primary wells.

- Wellhead

  The wellhead closes the geothermal wells at the surface. It usually includes a well silencer and an aluminum well housing containing the main wellhead valve as well as piping and smaller valves. The wellhead has negligible contributions to the environmental impacts of the construction phase; as such, the inventory developed by Karlsdóttir et al. (2015) can be generally applied.

- Collection pipelines

  Collection pipelines transport the geothermal fluid from the wells to the power plant. They are made of steel and insulated with mineral wool. They are sized based on the geothermal fluid mass flow and

NCGs. The material requirements for the collection pipelines can be estimated from the pipe diameter and thickness. It is suggested to use the inventory developed by Karlsdóttir et al. (2015) for the collection pipelines for the wellhead.

- Power plant

  The power plant includes the construction of several above-surface facilities and types of machinery required to enable and manage electricity generation from the geothermal fluid. Power plant construction has nonnegligible environmental impacts. However, since it is generally challenging to collect and develop a comprehensive inventory, in the first instance, it is suggested to use the inventory reported by Karlsdóttir et al. (2015) for single- and double-flash configurations. Note that geothermal power plants are not much different from conventional fossil fuel and nuclear power plants. The main distinction is that conventional plants require an additional component that generates thermal energy from fuel (e.g. coal, natural gas, or uranium).

- Heating station

When thermal energy is coproduced for district heating, the construction phase needs to include the heating station. This uses the remaining heat of the geothermal fluid to heat water that is sent to the district heating system. Like for the power plant, using the inventory reported by Karlsdóttir et al. (2015) is suggested as a first screening.

8.3.3.2 Operational phase

- Geothermal fluid pumping

  When the geothermal well is not self-flowing, the geothermal fluid is pumped out through downhole pumps. Their electricity consumption can be considerable due to the large flow rates often associated with geothermal systems. Typical electricity consumption values range between 0.5 and 2.5 kW/m^3 h (Frick et al., 2010; Lacirignola et al., 2014).

- Stimulation

  EGSs require hydraulic stimulation to create an engineered reservoir or enhance an existing one. Hydraulic stimulation aims at increasing far-well permeability by artificially creating new fractures in the rock with the injection of water at high pressure. A diesel-electric generating set is typically used to power the pumps, and the same considerations regarding diesel consumption for drilling apply to hydraulic stimulation. Besides hydraulic stimulation, chemical stimulation is also often applied at EGS sites and sometimes at conventional geothermal plants. Chemical stimulation improves near-well permeability by removing materials deposited in factures through dissolution in an acid substance. An inventory for hydraulic and chemical stimulation at the EGS site at Soultz-sous-Forêts can be found in Nami et al. (2008) and Schindler et al. (2010). The stimulation phase has been included in the operation phase, but it is not improper to incorporate it in the construction phase.

- Atmospheric emissions

  Atmospheric emissions of NCGs from dry-steam and single/multi-stage flash power plants are generally the largest source of climate change impacts. They may also significantly contribute to acidification and toxicity-related impacts. Therefore, CO_2 discharges must be included as a minimum. Discharges of CH_4 may also be considered, but it should be noticed that data for CH_4 are generally less available. Other NCGs of interest are H_2S, NH_3, and SO_2 for acidification and Hg, Pb, and As for toxicity categories. Hydrogen sulfide (H_2S) is an acidifying substance; however, a characterization factor is not currently provided in commonly used LCIA methods such as those developed by the Joint Research Centre (JRC, 2010) and the Institute of Environmental Sciences (CML) of Leiden University (Guinée et al., 2002). To estimate the impacts of H_2S from geothermal power plants, Karlsdottir et al. (2020) proposed using a factor of 1.88 to convert H_2S into SO_2 (for which characterization factors are available). The characterization factors for toxicity impacts of metals emissions feature significant uncertainties and should be treated with caution (Hauschild et al., 2013).

- Binary plant working fluid

  The amount of initial working fluid, as well as additional amounts required to make up unavoidable losses during operations, need to be included when electricity is generated by means of a binary cycle power plant. Indirect and direct emissions associated with the working fluid have nugatory contributions to all environmental categories except ozone depletion. Since inventory data for most working fluids are currently missing in commercial LCA databases, it is suggested to use a generic refrigerant as a proxy. The ecoinvent database (Wernet et al., 2016) includes R134 and assumes (1) a requirement of 300 kg per MW of installed capacity of the binary cycle plant and (2) 8% of yearly losses.

- Wastewater

The operational phase also needs to account for an appropriate treatment process of the geothermal fluid when this is not reinjected into the reservoir.

8.3.3.3 Maintenance phase

The maintenance phase needs to consider three activities; first, the need for additional wells drilling to make up for the decline in the flow rate of the geothermal fluid. The inventory shall be similar to that used in the construction phase. Sanyal (2004) assumes a harmonic decline of the wells' productivity to estimate the number of makeup wells required during the plant's lifetime. Second, the maintenance phase needs to account for equipment replacement such as downhole pumps, which can be quantified by means of replacement factors obtained as the ratio of the power plant lifetime by the specific lifetime of each equipment. Finally, the maintenance phase may also include consuming inhibitor chemicals that prevent scaling in the system.

8.3.3.4 End-of-life phase

Typically, LCA studies are performed when the geothermal plant is still in operation; therefore, end-of-life modeling requires the development of reasonable scenarios. The end-of-life phase includes closure of the wells and dismantling of above-surface facilities. The former typically fills the wells with cement and gravel and seals them with a concrete slab. This can be modeled in the absence of primary data by the corresponding activity in the ecoinvent database (Wernet et al., 2016). The latter requires considering the fate of dismantled materials, that is, landfilled, recycled, or incinerated. Notably, this can have significant effects on the results of the LCA. It is therefore suggested to developed different scenarios to test the sensitivity of end-of-life assumptions for above-surface facilities.

8.3.4 Life cycle inventory: handling multifunctional processes

In LCA, a process is defined as multifunctional when it delivers more than one function, that is, when it produces more than one good and/or provides more than one service. Combine heat-and-power plants are the most notable example in the geothermal sector. The extraction of valuable materials from the brine may represent an additional function for future geothermal plants.

Multifunctional processes pose a methodological challenge in LCA when the objective is the quantification of the environmental impacts associated with only one of the various functions, for example, electricity from a combined-heat-and-power plant. If, however, the objective of the study is to assess the environmental impacts of both electricity and heat generation, then there is no need for allocation. The ISO standards provide a practical stepwise procedure for handling multifunctional processes consisting of three steps (ISO, 2006b). It must be noted that the standards do not (yet) distinguish between different LCA "modes" and thus recommend using the same procedure for both attributional and consequential approaches. However, many maintain that the appropriate allocation method depends on the mode of LCA (Ekvall and Finnveden, 2001; Finnveden et al., 2009; JRC, 2010; Weidema, 2003).

The first step in the ISO procedure suggests subdividing the process into several subprocesses that can only be attributed to one function. This can only be partially applied to the geothermal sector. Construction and dismantling of the power plant and the heating station can be allocated to one function only, respectively, electricity and heat generation. Operational impacts such as discharges of NCGs, water consumption for district heating, and losses of the working fluid can also be allocated to one function only. However, construction and closure of geothermal wells and collection pipelines cannot be allocated to one function only; thus, they need to be treated by "partitioning."

The second step, that is, system expansion, is primarily used for comparative assessment when the compared product system does not deliver the same functions. For instance, when comparing combined-heat-and-power geothermal plants with other power plant types that generate only electricity like onshore/offshore wind farms or solar photovoltaic. In this case, the compared product system is expanded to include the additional heat generation function represented by the most likely technology (consequential approach) or a mix of technologies (attributional approach). System expansion is mathematically equivalent to subtracting from the studied product system the environmental impacts of the alternative way to provide the additional function. This approach is known as "crediting" (Bjorn et al., 2018).

The third step consists of partitioning the inputs and outputs among the different functions, using underlying physical relationships or, when this is not possible, other relationships like the economic value of the products. Several partitioning strategies have been developed.

The "alternative generation method" is recommended for using the Environmental Product Declaration for heat and power cogeneration (Environdec, 2019). The most common approaches used in the literature use either energy or exergy content or the economic value. Paulillo et al. (2019b) showed the difference between these strategies when estimating the environmental footprint of the Hellisheiði power plant in Iceland. Tereshchenko and Nord (2015) compared seven partitioning methods (excluding economic). They found that different methods can lead to substantial differences in the LCA results, ranging from 28% of the impacts allocated to electricity by the alternative generation method to 94% by the 200% method. They also found that the choice of the allocation method is more important than other parameters like the size of the plant, properties of the distribution network, and plant technology. The need to develop a consensus partitioning method is evident. In the absence, it is recommended to use a method that (1) is not excessively sensitive to other parameters and (2) gives intermediate partitioning factors. According to the study by Tereshchenko and Nord (2015), these are exergy, energy, and the power bonus method.

8.3.5 Life cycle impact assessment

Nearly all LCA studies on geothermal power include climate change impacts, while a smaller number include other impact categories and even fewer include all categories within an LCIA method, that is, a predefined set of impact category models and indicators developed by specific organizations like the Institute of Environmental Sciences at Leiden University (CML) and the Joint Research Centre (JRC) of the European Commission. This is because the objective of most studies is to assess the potential of geothermal energy to support carbon reduction goals and climate change policies. However, including one or a minimal subset of impact categories is limiting. It precludes the identification of trade-offs among different environmental issues—indeed, one of the key features of LCA.

This section attempts to recommend appropriate LCIA methods and impact categories for the geothermal power sector. Two LCIA methods have been used for assessing the environmental impacts of geothermal power. The method developed by the JRC—and, for example, used by Paulillo et al. (2019a, 2019b)—is based on a systematic review of available approaches for impact assessment (Hauschild et al., 2013) and is generally recommended for use in Europe. The method developed by the CML institute (Guinée et al., 2002) is perhaps one of the most used LCIA methods, for example, in the geothermal power sector by Bravi and Basosi (2014), Karlsdottir et al. (2020), Martín-Gamboa et al. (2015). It is recommended to use either of these methods unless a method with a strong national focus—such as Lime in Japan or Traci in the US—is required.

In the absence of a better approach, recommendations for selecting the environmental impact categories are based on their normalized impacts, as reported by (Paulillo et al., 2019a, 2020b) for a single- and double-flash and an enhanced geothermal plant. The most critical categories—those with the highest normalized values—are environmental and human toxicity (mainly associated with indirect impacts from casing steel), particulate matter and photochemical ozone formation (due to indirect and direct emissions from diesel), and climate change for dry-steam and single- and multistage flash plants only (from direct operational emissions of NCGs). It must be noted that selecting environmental categories solely on the basis of their normalized impacts may be open to criticism. The objective of normalization is to identify the environmental categories that contribute the most to the environmental footprint of a reference system, but contributing the most is not necessarily equivalent to being the most important. A high normalized impact does not identify a critical category when the impact of the reference system in that category is low. For instance, when compared to a threshold like the Planetary Boundaries (Steffen et al., 2015). In this case, a high normalized value simply indicates that the product system represents a large contributor to the impact of the reference system in that category.

8.3.6 Life cycle interpretation: reporting LCA results

This section focuses on one of the aspects of life cycle interpretation: how to report LCA results. For geothermal power, it is recommended to report the results in terms of (1) distribution of the impacts by phase (i.e., construction, operation, maintenance, and end-of-life), (2) distribution between direct and indirect impacts, and (3) individual emissions/wastes or activities that explain a large portion of the environmental impacts. The distribution of the impacts by phase is useful since it enables understanding the relative importance of the parameters used in modeling the product system and thus in drawing significant conclusions. For instance, in the case of geothermal power, the construction phase is the largest contributor to most impact categories. This implies that increasing the lifetime of the plant would substantially increase the environmental performance. Reporting direct and indirect impacts is especially useful to understand the importance of atmospheric discharges from the plants primarily to the environmental category climate change and acidification. Finally, identifying individual emissions/ wastes and activities that have significant contributions to the environmental impacts is essential. It represents the basis for drawing conclusions and recommendations for improvements. It is also paramount to accompany LCA results with the specific parameters used to model the geothermal resource and the power plant for transparency and reproducibility purposes.

8.4 Case studies

This section provides a few examples of results from LCA studies conducted on the geothermal power sector. The objective is to demonstrate how LCA can be used to quantify and compare the environmental footprint, identify approaches that could improve the environmental performance, and ultimately provide information to decision-makers. The focus is on two specific applications: hot-spot analysis and comparative analysis.

8.4.1 Hot-spot analysis

Hot-spot (or contribution) analysis aims at identifying the dominant sources of environmental impacts of a product system (i.e., the hot-spots) in terms of, for instance, phases, activities or individual emissions, resources, and wastes. Fig. 8.6 reports the hot-spot analysis conducted for the Hellisheiði field site, a combined-heat-and-power plant located in Iceland with an installed capacity of 303 MW for electricity and 133 MW for hot water (Paulillo et al., 2019b). The figure includes two charts, which refer to the entire product system (on the left) and the construction of wells (on the right). The charts report a selected number of impact categories based on the ILCD method (JRC, 2010). The hot-spot analysis reveals that the construction phase represents the dominant source of impacts in all categories but climate change and ecotoxicity. The former is dominated by operational atmospheric emissions of noncondensable greenhouse gases, primarily CO_2, while the latter by the end-of-life phase, primarily from the disposal of the copper used in the power plant. The results obtained from the LCA analysis enable drawing some crucial recommendations. First, increasing the power plant's lifetime and that of the geothermal reservoir is an effective strategy for reducing the environmental impacts of geothermal power production. Second, a carbon capture system is likely to substantially reduce

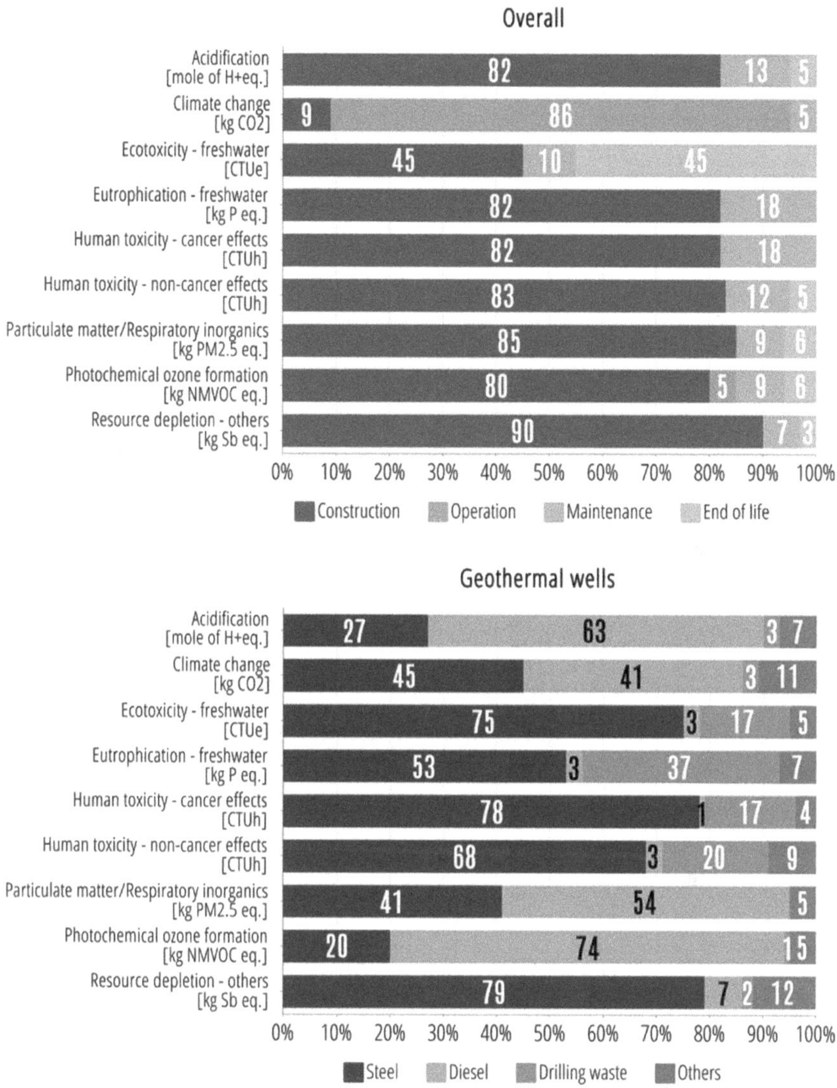

Figure 8.6 Hot-spot analysis for the Hellisheiði geothermal power plant. Source: *(above) Adapted from Paulillo, A., Striolo, A.A., Lettieri, P., 2019b. The environmental impacts and the carbon intensity of geothermal energy: a case study on the Hellisheiði plant. Environment International 133(Pt B). doi:10.1016/j.envint.2019.105226. (below) Paulillo, A., Striolo, A., Lettieri, P., 2019a. Data on the environmental impacts and the carbon intensity of geothermal energy: a case study on the Hellisheiði plant. Data in Brief. 27, 104771.*

the power operations' carbon footprint at the Hellisheiði plant. Note that, at present, ~ 30% of CO_2 emissions are captured and stored into basaltic rocks via an innovative technology known as CO_2 mineralization (Karlsdottir et al., 2020).

The construction of the geothermal wells represents the largest contributor to the environmental impact during the construction phase. As such, it needs to be analyzed in detail. This is the role of Fig. 8.6, which shows that the environmental impacts associated with the construction of geothermal wells originate from the indirect emissions due to steel production, which is used as casing for the wells, and to diesel production and its use on-site to generate electricity for powering the drilling rig. These results also enable drawing some recommendations. First, using electricity from the grid instead of diesel to power the drilling rig is likely to reduce the environmental impacts of a number of categories, including acidification, photochemical ozone formation, and particulate matter. Indeed, the effects of using electricity depend on the specific local or national grid mix. Second, increasing recycled steel content is also likely to reduce the impact in the toxicity and eutrophication categories.

8.4.2 Comparative analysis

Comparative analysis is conducted to measure the environmental performance of a product system relative to other systems providing similar products or services. Geothermal energy is thus typically compared with other power generation technologies, whether renewables or not. For example, it might be instructive to compare the environmental footprint of geothermal energy production against that connected with onshore/offshore wind, solar photovoltaic, boiling and pressure water nuclear reactors, and fossil fuel-powered plants for the provision of electricity. Geothermal energy can also be compared with other technologies for the provision of thermal energy. For instance, it can be compared with electric or natural gas boilers or heat pumps for the provision of space heating. Two examples of comparative analysis are provided below: one focusing on the carbon footprint of geothermal technologies, and the other comparing the environmental footprint of geothermal energy against that expected for other energy technologies.

8.4.2.1 Carbon footprint

Carbon footprint-based comparisons are the most common type of comparison for two reasons. Firstly, most studies on power generation technologies only focus on quantifying climate change impacts. This is because the objective is to understand the potential of geothermal energy to decarbonize the power generation sector. Secondly, there is only one accepted model for assessing the impact of greenhouse gases on global warming (developed by the IPCC). By contrast, there are typically several impact assessment models for each environmental category. This implies that comparing carbon footprint across different studies is always possible, but not for other categories.

Fig. 8.7 reports the carbon footprint (in terms of g CO_2 equiv. per kWh) of different geothermal technologies, including dry-steam, single and double-flash, binary cycle for conventional and "engineered" (i.e., EGS) reservoirs, obtained from Paulillo et al. (2019a). The data suggest that direct operational emissions of NCGs, primarily CO_2 and CH_4, drive the carbon footprint of geothermal power technologies. That is why dry-steam and single-flash power plants exhibit the most significant carbon footprint, and binary cycle plants have the lowest. However, these results must be interpreted with caution. Firstly, since there are more than five data points only for the binary (EGS) technology, and secondly, because the carbon footprint of geothermal power is exceptionally location-specific. It depends on the specific characteristics of the geothermal fluid and its composition. In particular, the data for dry-steam and single-flash plants are based on geothermal plants in the region of Tuscany, Italy, known for having a large concentration of methane dissolved in the geothermal fluid.

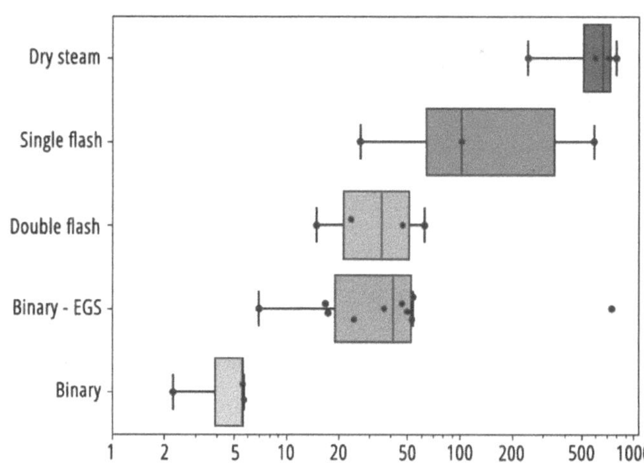

Figure 8.7 Carbon footprint of geothermal power technologies. Source: *Adapted from Paulillo, A., Striolo, A., Lettieri, P., 2019a. Data on the environmental impacts and the carbon intensity of geothermal energy: a case study on the Hellisheiði plant. Data in Brief. 27, 104771.*

8.4.2.2 Multiple environmental categories

When possible, it is preferable to compare the environmental performance of different technologies across multiple environmental categories. This enables the identification of trade-offs. Environmental superiority can only be declared if one technology outperforms another across all environmental categories; this rarely occurs. In most cases, trade-offs need to be considered in the decision-making process. Fig. 8.8 reports the expected environmental impacts of electricity generation from the United Downs Deep Geothermal Power (UDDGP) plant, constructed in Cornwall, United Kingdom. This is compared with the environmental impacts of four energy technologies—offshore wind, solar photovoltaic, natural gas-powered combined cycle gas turbine, and nuclear-pressurized water reactors—which are likely to play an important role in the future energy system of the UK (Paulillo et al., 2020a). The comparison exposes

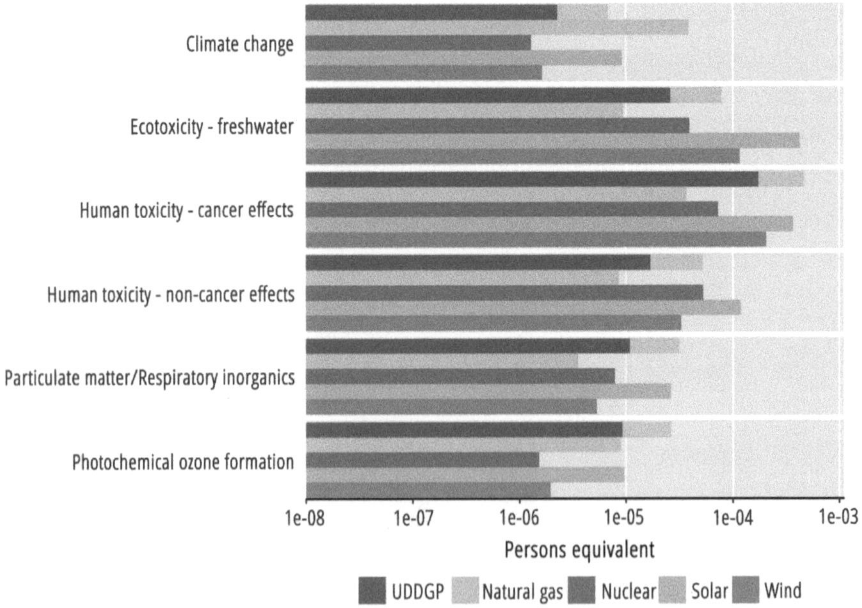

Figure 8.8 Comparative analysis of normalized environmental impacts associated with electricity production from the United Downs Deep Geothermal Power (UDGGP) plant, offshore wind, solar photovoltaic, nuclear-pressurized water reactors, and combined cycle gas turbine. Source: *Adapted from Paulillo, A., Cotton, L., Law, R., Striolo, A., Lettieri, P., 2020a. Geothermal energy in the UK: the life-cycle environmental impacts of electricity production from the United Downs Deep Geothermal Power project. Journal of Cleaner Production. doi:10.1016/j.jclepro.2019.11941.*

numerous trade-offs. The UDDGP plant has good performance in the climate change and toxicity categories. However, it features the highest or second-highest impact in photochemical ozone formation and particulate matter categories. Therefore, the UDDGP plant, in its current design, represents a good option to reduce the carbon intensity of the UK grid mix. However, the technology is not environmentally superior to other available energy technologies.

Acknowledgments

This work is part of the S4CE collaborative project, which has received funding from the European Union's Horizon 2020 research and innovation program under grant agreement number 764810.

References

Baldacci, A., Mannari, M., Sansone, F., 2005. Greening of geothermal power : an innovative technology for abatement of hydrogen sulphide and mercury emission, in proceedings of the World Geothermal Congress. Antalya, Turkey, 2005.

Baumann, H., Tillman, A.M., 2004. The Hitch Hiker's Guide to LCA: An Orientation in Life Cycle Assessment Methodology and Application. Studentlitteratur, Lund, Sweden, Sweden. Available from: https://doi.org/10.1065/lca2006.02.008.

Bertani, R., 2016. Geothermal power generation in the world 2010–2014 update report. Geothermics 60, 31–43. Available from: https://doi.org/10.1016/j.geothermics.2015.11.003.

Bjorn, A., Moltesen, A., Laurent, A., Owsianiak, M., Corona, A., Birkved, M., et al., 2018. Life cycle inventory analysis. In: Hauschild, M.Z., Rosenbaum, R.K., Olsen, S. I. (Eds.), Life Cycle Assessment. Theory and Practice. Springer International Publishing.

Bravi, M., Basosi, R., 2014. Environmental impact of electricity from selected geothermal power plants in Italy. Journal of Cleaner Production 66, 301–308. Available from: https://doi.org/10.1016/j.jclepro.2013.11.015.

Breede, K., Dzebisashvili, K., Liu, X., Falcone, G., 2013. A systematic review of enhanced (or engineered) geothermal systems: past, present and future. Geothermal Energy 1. Available from: https://doi.org/10.1186/2195-9706-1-4.

Cataldi, R., Hogdson, S.F., Lund, J.W., 1999. Stories from a Heated Earth: Our Geothermal Heritage. Geothermal Resources Council, Davis.

DiPippo, R., 2016a. Geothermal Power Plants: Principles, Applications, Case Studies and Environmental Impact, 4th ed. Butterworth-Heinemann.

DiPippo, R., 2016b. Geothermal Power Generation: Developments and Innovation. Woodhead Publishing.

DiPippo, R., 2016c. Dry-steam power plants. In: Di Pippo, R. (Ed.), Geothermal Power Plants: Principles, Applications, Case Studies and Environmental Impact. Butterworth-Heinemann.

DiPippo, R., 2016d. Single-flash steam power plants. In: Di Pippo, R. (Ed.), Geothermal Power Plants: Principles, Applications, Case Studies and Environmental Impact. Butterworth-Heinemann.

DiPippo, R., 2016e. Double- and triple-flash steam power plants. In: Di Pippo, R. (Ed.), Geothermal Power Plants: Principles, Applications, Case Studies and Environmental Impact. Butterworth-Heinemann.
DiPippo, R., 2016f. Binary cycle power plants. In: Di Pippo, R. (Ed.), Geothermal Power Plants: Principles, Applications, Case Studies and Environmental Impact. Butterworth-Heinemann.
Dunstall, M., Foster, B., 1998. Geothermal greenhouses at Kawerau. GHG Bulletin.
Ekvall, T., Finnveden, G., 2001. Allocation in ISO 14041: a critical review. Journal of Cleaner Production 9, 197–208. Available from: https://doi.org/10.1016/S0959-6526(00)00052-4.
Environdec, 2019. Product Group Classification: UN CPC 171 and 173. Electricity, steam and hot/cold water generation and distribution.
Finnveden, G., Hauschild, M.Z., Ekvall, T., Guinee, J.B., Heijungs, R., Hellweg, S., et al., 2009. Recent developments in Life Cycle Assessment. Journal of Environmental Management 91, 1–21. Available from: https://doi.org/10.1016/j.jenvman.2009.06.018.
Frick, S., Kaltschmitt, M., Schröder, G., 2010. Life cycle assessment of geothermal binary power plants using enhanced low-temperature reservoirs. Energy 35, 2281–2294. Available from: https://doi.org/10.1016/j.energy.2010.02.016.
Gando, A., Gando, Y., Ichimura, K., Ikeda, H., Inoue, K., Kibe, Y., et al., 2011. Partial radiogenic heat model for Earth revealed by geoneutrino measurements. Nature Geoscience 4, 647–651. Available from: https://doi.org/10.1038/ngeo1205.
Guinée, J.B., Gorrée, M., Heijungs, R., Huppes, G., Kleijn, R., de Koning, A., et al., 2002. Handbook on Life Cycle Assessment. Operational Guide to the ISO Standards. Kluwer Academic Publishers, Dordrecht.
Harvey, W., Wallace, K., 2016. Flash steam geothermal energy conversion systems: single-, double-, and triple-flash and combined-cycle plants. In: Di Pippo, R. (Ed.), Geothermal Power Generation: Developments and Innovation. Woodhead Publishing.
Hauschild, M.Z., Goedkoop, M., Guinée, J.B., Heijungs, R., Huijbregts, M.A.J., Jolliet, O., et al., 2013. Identifying best existing practice for characterization modeling in life cycle impact assessment. The International Journal of Life Cycle Assessment 18, 683–697. Available from: https://doi.org/10.1007/s11367-012-0489-5.
Hauschild, M.Z., Rosenbaum, R.K., Olsen, S.I., 2018. Life Cycle Assessment: Theory and Practice. Springer International Publishing. Available from: https://doi.org/10.1007/978-3-319-56475-3.
IEA, 2019a. Renewables 2019. Market Analysis and Forecast from 2019 to 2024. IEA: San Francisco, USA.
IEA, 2019b. World Energy Outlook 2019. San Francisco, CA: IEA.
IFC, 2013. Success of Geothermal Wells: A Global Study. Washington, DC: IFC.
IGA, 2015. Geothermal power database [www document]. https://www.geothermal-energy.org/explore/our-databases/geothermal-power-database/#electricity-generation-by-plant (accessed 24.07.19.).
IRENA, 2019. Renewable Power Generation Costs in 2018. https://doi.org/10.1007/SpringerReference_7300.
ISO, 2006a. Environmental management - life cycle assessment - principles and framework. EN ISO 14040:2006. ISO.
ISO, 2006b. Environmental management - life cycle assessment - requirements and guidelines. EN ISO 14044:2006. ISO.
JRC, 2010. ILCD Handbook - general guide for Life Cycle Assessment - detailed guidance. https://doi.org/10.2788/38479.

Karlsdottir, M.R., Heinonen, J., Palsson, H., Palsson, O.P., 2020. Life cycle assessment of a geothermal combined heat and power plant based on high temperature utilization. Geothermics 84, 101727. Available from: https://doi.org/10.1016/j.geothermics.2019.101727.

Karlsdóttir, M.R., Pálsson, Ó.P., Pálsson, H., Maya-Drysdale, L., 2015. Life cycle inventory of a flash geothermal combined heat and power plant located in Iceland. The International Journal of Life Cycle Assessment 20, 503–519. Available from: https://doi.org/10.1007/s11367-014-0842-y.

Lacirignola, M., Blanc, I., 2013. Environmental analysis of practical design options for enhanced geothermal systems (EGS) through life-cycle assessment. Renewable Energy 50, 901–914. Available from: https://doi.org/10.1016/j.renene.2012.08.005.

Lacirignola, M., Meany, B.H., Padey, P., Blanc, I., 2014. A simplified model for the estimation of life-cycle greenhouse gas emissions of enhanced geothermal systems. Geothermal Energy 2, 1–19. Available from: https://doi.org/10.1186/s40517-014-0008-y.

Lu, S.M., 2018. A global review of enhanced geothermal system (EGS). R enewable and Sustainable Energy Reviews 81, 2902–2921. Available from: https://doi.org/10.1016/j.rser.2017.06.097.

Lund, J., 2007. Development and utilization of geothermal resources. In: The Proceedings of ISES World Congress. Springer, Berlin.

Marchand, M., Blanc, I., Marquand, A., Beylot, A., Bezelgues-Courtade, S., Traineau, H., 2015. Life Cycle Assessment of high temperature geothermal energy systems. In: Proceedings of World Geothermal Congress 2015. HAL.

Martín-Gamboa, M., Iribarren, D., Dufour, J., 2015. On the environmental suitability of high- and low-enthalpy geothermal systems. Geothermics 53, 27–37. Available from: https://doi.org/10.1016/j.geothermics.2014.03.012.

Menberg, K., Pfister, S., Blum, P., Bayer, P., 2016. A matter of meters: state of the art in the life cycle assessment of enhanced geothermal systems. Energy & Environmental Science 9, 2720–2743. Available from: https://doi.org/10.1039/c6ee01043a.

Mines, G., 2016. Binary geothermal energy conversion systems: basic Rankine, dualepressure, and dualefluid cycles. In: Di Pippo, R. (Ed.), Geothermal Power Generation: Developments and Innovation. Woodhead Publishing.

MIT, 2006. The future of geothermal energy. Impact of enhanced geothermal systems (EGS) on the United States in the 21st Century. MIT, MA, USA.

Nami, P., Schellschmidt, R., Schindler, M., Tischner, T., 2008. Chemical stimulation operations for reservoir development of the deep crystalline hdr/egs system At Soultz-Sous-Forêts (France). In: Proceedings of the 32nd workshop on geothermal reservoir engineering, Stanford University Press, Stanford, CA, pp. 325–333.

Parri, R., Lazzeri, F., 2016. Larderello: 100 years of geothermal power plant evolution in Italy. In: Di Pippo, R. (Ed.), Geothermal Power Generation: Developments and Innovation. Woodhead Publishing.

Paulillo, A., Cotton, L., Law, R., Striolo, A., Lettieri, P., 2020a. Geothermal energy in the UK: the life-cycle environmental impacts of electricity production from the United Downs Deep Geothermal Power project. Journal of Cleaner Production. Available from: https://doi.org/10.1016/j.jclepro.2019.119410.

Paulillo, A., Cotton, L., Law, R., Striolo, A., Lettieri, P., 2020b. Life-cycle inventory and impacts on electricity production at the United Downs Deep Geothermal Power project in the UK. Data in Brief 29, 105117.

Paulillo, A., Striolo, A., Lettieri, P., 2019a. Data on the environmental impacts and the carbon intensity of geothermal energy: a case study on the Hellisheiði plant. Data in Brief 27, 104771.

Paulillo, A., Striolo, A.A., Lettieri, P., 2019b. The environmental impacts and the carbon intensity of geothermal energy: a case study on the Hellisheiði plant. Environment International 133 (Pt B). Available from: https://doi.org/10.1016/j.envint.2019.105226.

Phair, K., 2016. Direct steam geothermal energy conversion systems: dry steam and superheated steam plants. In: Di Pippo, R. (Ed.), Geothermal Power Generation: Developments and Innovation. Woodhead Publishing.

Pratiwi, A., Ravier, G., Genter, A., 2018. Life-cycle climate-change impact assessment of enhanced geothermal system plants in the Upper Rhine Valley. Geothermics 75, 26–39. Available from: https://doi.org/10.1016/j.geothermics.2018.03.012.

Rybach, L., 2007. Geothermal sustainability. GHC Bulletin 1–7. Available from: https://doi.org/10.1046/j.0956-540x.2001.01558.x.

Rybach, L., 2003. Geothermal energy: sustainability and the environment. Geothermics 32, 463–470. Available from: https://doi.org/10.1016/S0375-6505(03)00057-9.

Santoyo, E., Verma, S.P., Nieva, D., Portugal, E., 1991. Variability in the gas phase composition of fluids discharged from Los Azufres geothermal field, Mexico. Journal of Volcanology Geothermal Research . Available from: https://doi.org/10.1016/0377-0273(91)90107-B.

Sanyal, S.K., 2004. Cost of geothermal power and factors that affect it. Proceedings of the Twenty-Ninth Workshop on Geothermal Reservoir Engineering. Stanford University Press, Stanford, CA, pp. 24–29.

Schindler, M., Baumgärtner, J., Gandy, T., Hauffe, P., Hettkamp, T., Menzel, H., et al., 2010. Successful hydraulic stimulation techniques for electric power production in the Upper Rhine Graben. Central Europe. Proceedings of the World Geotherm. Congress World Geotherm, Bali, pp. 1–7.

Smith, Y., Kumar, P., McLennan, J., 2017. On the extraction of rare earth elements from geothermal brines. Resources 6. Available from: https://doi.org/10.3390/resources6030039.

Steffen, W., Richardson, K., Rockstrom, J., Cornell, S.E., Fetzer, I., Bennett, E.M., et al., 2015. Planetary boundaries: guiding human development on a changing planet. Science 347. Available from: https://doi.org/10.1126/science.1259855 (80-.).

Tereshchenko, T., Nord, N., 2015. Uncertainty of the allocation factors of heat and electricity production of combined cycle power plant. Applied Thermal Engineering 76, 410–422. Available from: https://doi.org/10.1016/j.applthermaleng.2014.11.019.

Weidema, B., 2003. Market Information in Life Cycle Assessment Environment Project No. 863. Danish Environment Protection Agency.

Wernet, G., Bauer, C., Steubing, B., Reinhard, J., Moreno-Ruiz, E., Weidema, B., 2016. The ecoinvent database version 3 (part I): overview and methodology. The International Journal of Life Cycle Assessment 21, 1218–1230. Available from: https://doi.org/10.1007/s11367-016-1087-8.

White, D., 1973. Characteristics of geothermal resources. In: Kruger, P., Otte, C. (Eds.), Geothermal Energy: Resources, Production, Stimulation. Stanford University Press, Stanford, CA.

White, D., Wiliams, D., 1975. Assessment of geothermal resources of the United States. US Department of the Interior, Geological Survey.

Wild, M., Folini, D., Schär, C., Loeb, N., Dutton, E.G., König-Langlo, G., 2013. The global energy balance from a surface perspective. Climate Dynamics 40, 3107–3134. Available from: https://doi.org/10.1007/s00382-012-1569-8.

Zarrouk, S.J., Moon, H., 2014. Efficiency of geothermal power plants: a worldwide review. Geothermics 51, 142–153. Available from: https://doi.org/10.1016/j.geothermics.2013.11.001.

SECTION C

CHAPTER 9

Environmental impact assessment of hydropower stations

M. A. Parvez Mahmud[1] and Nahin Tasmin[2]
[1]School of Engineering, Deakin University, Geelong, VIC, Australia
[2]Department of Mechanical Engineering, Rajshahi University of Engineering & Technology, Kazla, Rajshahi, Bangladesh

Contents

9.1 Introduction

Global energy demand is increasing rapidly, owing to the rising world population along with the industries. To fulfill the expanding energy demand, the rate of conventional power generation is also increasing. Accordingly, a huge amount of harmful greenhouse gas (GHG) emissions is getting released from these fossil-fuel-based power plants (Das et al., 2011; Zou et al., 2016; Farjana et al., 2018). For that reason, the world climate is also facing adverse change (Capellán-Pérez et al., 2016; Mahmud and Farjana, 2012; Armaroli and Balzani, 2011; Parvez Mahmud et al., 2018; Ward and Inderwildi, 2013). People are becoming more responsive about the ecosystem to avoid the future inevitable threats of high-risk emissions to the environment and are willing to lower the levels of global CO_2 emissions for an enhanced life comfort. Therefore, renewable energy production systems are gradually getting considerable attention to rescue the environment from pollution and narrow the scale of conventional power-generation systems (Turconi et al., 2013; Atilgan and

Environmental Assessment of Renewable Energy Conversion Technologies
DOI: https://doi.org/10.1016/B978-0-12-817111-0.00005-X

Azapagic, 2015; Mahmud et al., 2016; Santoyo-Castelazo et al., 2011; Espinosa et al., 2014; Parvez Mahmud and Farjana, 2012; Périlhon et al., 2012; Botelho et al., 2017; Uddin and Kumar, 2014; Gong et al., 2015). It is very much certain that the usage of renewable energy sources in power plants will reach a peak value in the upcoming years. Hydropower is taken as the cleanest energy source for power generation amid other renewable energy technologies (RET) (Botelho et al., 2017; de Miranda Ribeiro and Da Silva, 2010; Li et al., 2017). The accessibility of resources in European alpine and nonalpine areas has frequently persuaded the installation of power plants in these regions (Atilgan and Azapagic, 2015; Wagner et al., 2017; Lazzaro and Botter, 2015).

Nevertheless, hydropower plants also release dangerous emissions as obtained from previous research (de Miranda Ribeiro and Da Silva, 2010; Li et al., 2017). Consequently, it is very much reasonable to examine the pollution caused by hydropower plants. However, very limited research has been conducted to assess the hazardous emissions from the preparation of the plant route. A comparative analysis to evaluate the perilous environmental impacts from hydropower plants in Portugal was made by Botelho et al. (2017) They were marking up critical issues such as that plants and animals are the most usual victims of the impacts. Their research did not include the whole life cycle assessment (LCA) of the plants. de Miranda Ribeiro and Da Silva (2010) studied the life cycle inventory (LCI) of a Brazilian hydroelectric power plant, but they did not employ any global database such as ecoinvent. Li et al. (2017) reported the carbon footprints of two large Chinese hydro-projects via LCA in line with the International Organization for Standards (ISO) 14067. However, their analysis excluded metal-based emissions and relevant detailed impacts on some indicators such as human health, ecosystem, resources, and climate changes. Nevertheless, to the authors' best knowledge, no research has been conducted to realize the comparative effects of hydropower plants in alpine and nonalpine regions. Hence, it is necessary to estimate the impacts by a systematic approach considering all input and output flows during the whole lifespan and the end-of-life waste management schemes of the plants. Therefore, it is essential to measure the impacts by a systematic approach considering all input and output flows during the whole lifespan and the end-of-life waste management schemes of the plants. The objective of this research is to quantify the risk effects of alpine and nonalpine hydropower plants using the LCA approach. LCA is a widely recognized technique for estimating the ecological effects of a power generation plant throughout its lifespan from raw-materials extraction to processing, transport, operation, and end-of-life disposal (Uddin and Kumar, 2014; Gong et al., 2015). The SimaPro software is used as it contains the latest LCA-based methods (Marsmann and Schiburr, 1993), and the ecoinvent database is applied to evaluate the impacts by systematic LCA approach. The reason behind SimaPro software and ecoinvent database application is their global acceptance (Pascual-González et al., 2016; Frischknecht and Rebitzer, 2005). The obtainable impacts fall under 14 indicators: carcinogens, noncarcinogens, respiratory inorganics, ionizing radiation, ozone-layer depletion, respiratory organics, aquatic ecotoxicity, terrestrial ecotoxicity, terrestrial acid, land occupation, aquatic acidification, aquatic eutrophication, global warming, non-renewable energy, and mineral extraction. Methods employed in evaluating these effects are the ReCiPe and Impact 2002 + , which further fall into four prime indicators: resources, climate change, ecosystem quality, and human health. Metal- and gas-based impending emissions are estimated by the Ecopoints 97 method.

Besides, GHG (carbon dioxide, methane, nitrous oxide, etc.) emissions into the air are assessed by the intergovernmental panel on climate change (IPCC) method.

Hence, the key contributions of this work can be outlined as presented below:

- Conducting investigation in evaluating the environmental pollution from the existing hydropower plants in Europe.
- Designing a dignified LCI to estimate the environmental hazards connected to the ecosystem, climate change, resources, and human health for both alpine and nonalpine hydropower plants of Europe.
- Carrying out a step-by-step LCA analysis to measure the GHG emissions of both plants.
- Finishing the uncertainty analysis for both hydropower plants.

This research work is distinctive due to the employment of various standardized methodologies in evaluating the environmental impacts of alpine and nonalpine hydropower plants in Europe, taking into account both the cradle-to-gate and the cradle-to-grave approaches.

In the light of the preceding, the rest of the paper is organized as follows. In Section 9.2, a review of the LCA method is highlighted. Section 9.3 discusses the outcomes of the environmental profiles of the plants, metal- and gas-based emission evaluation of the plants, greenhouse-gas emission estimation of the plants, and uncertainty analysis of the plants. Finally, in Section 9.4, the conclusions on the research outcomes and future recommendations are discussed.

9.2 Materials and methods

The objective of the present research work was to investigate the environmental impact of hydropower plants of alpine and nonalpine areas of Europe. The broadly acknowledged LCA method was employed to obtain the effects. The LCA method is also a well-built tool for scanning the carbon footprints of any product, unit, or system (Heijungs and Guinée, 2012; Lewandowska et al., 2011). The LCA has frequently been used to calculate the ecological impacts and sustainability checks (Keoleian, 1993; Pesso, 1993; de Larderel, 1993; de Haes, 1993), generally conducted by following the ISO standards of 14040:2006 and 14044:2006 (Pryshlakivsky and Searcy, 2013; Finkbeiner et al., 2006). The inventories obtained from the ecoinvent database for the alpine- and nonalpine-based hydropower plants in Europe were the scope of this research. The LCA was performed with the use of SimaPro software, while the selected analysis methods were ReCiPe, Impact 2002 + , Ecopoints, and IPCC. All these methods are vastly used for life cycle environmental impact assessment associated with energy generation technologies (Stavropoulos et al., 2016; Curran, 2013; Židonienė and Kruopienė, 2015). The geographical coverage associated with this LCA analysis is both cradle-to-gate (midpoint) and cradle-to-grave (endpoint). LCA was accomplished by maintaining the four fundamental stages: defining scope and objective, LCI analysis, life cycle environmental impact evaluation, and impact outcome interpretation. The prime objective in the first step of LCA was the relative assessment of the environmental threats of the alpine- and nonalpine-based hydropower plants. This step was of help in identifying the most suitable hydropower plants' location in Europe from an environmental viewpoint. The LCA was done considering cradle-to-gate and cradle-to-grave aspects for both of the plants. Hence, the sum inputs and outputs obtained from the database of the entire lifetime of the power plants were taken into account to evaluate the impacts. The rate of energy and material flow in different stages of the considered hydropower plants are illustrated in Figs. 9.1 and 9.2. For the rate reference flow and functional unit, 1MJ of energy production was taken into consideration (Heijungs and Guinée, 2012; Lewandowska et al., 2011).

In the second step of the LCA assessment, an LCI was built, including all of the input−output materials, energies, and emissions. An LCA system boundary was used for this analysis for this objective, as presented in Fig. 9.3. To construct, the LCI data was assembled from data by the

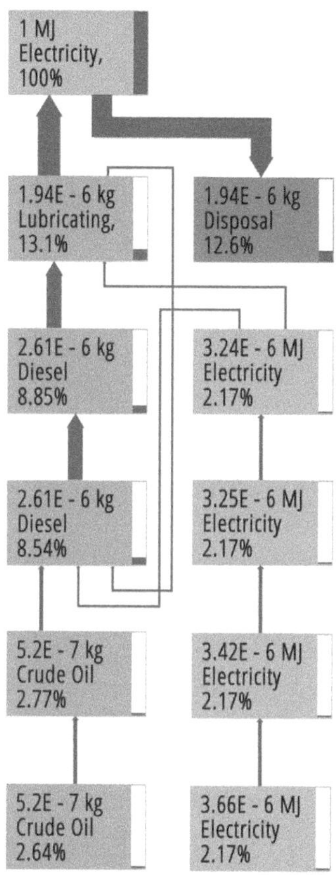

Figure 9.1 Material flow sheet for 1 MJ of hydropower generation in an alpine region.

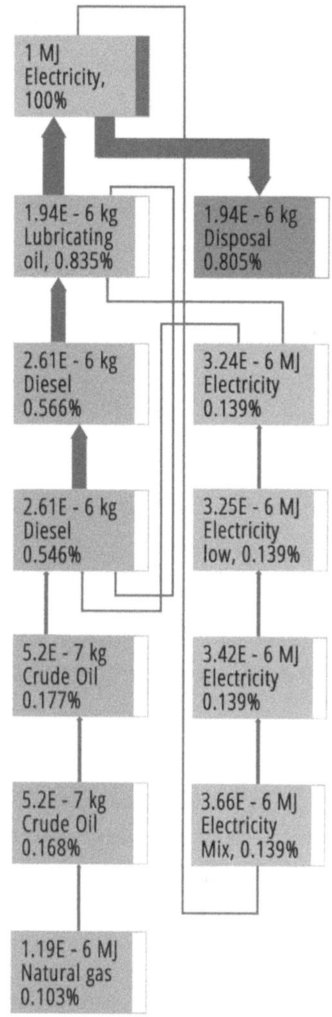

Figure 9.2 Material flow sheet for 1 MJ of hydropower generation in a nonalpine region.

acknowledged ecoinvent database (Pascual-González et al., 2016; Frischknecht and Rebitzer, 2005). The LCA boundary system examined in this process can be depicted in Fig. 9.3. As presented, the total inputs of a plant were raw materials, water, non-renewable energy, organic and inorganic chemicals, transportation, and other available resources at the specific plant location. The output is 1 MJ of electricity production and

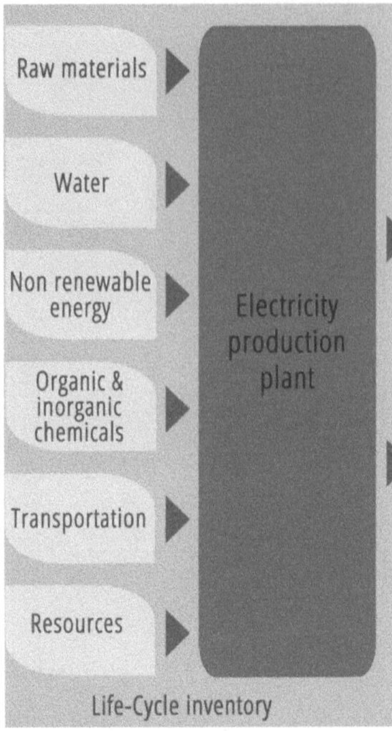

Figure 9.3 Life cycle assessment system boundary.

waste emissions. Table 9.1 groups the inputs from nature and the technosphere and outputs to air and technosphere by the considered plants collected from the LCI datasets. As depicted, the database for input portions is nearly similar for both plants. It shows that the input amounts were almost identical for both plants. There is the same amount of oil disposal for both plants; however, in terms of biogenic methane emissions, hydroelectricity generation plants in nonalpine-regions were overridden that of alpine regions.

The third step of the LCA process was the LCIA, which was performed by following ISO 10442. In this step, emissions and input variables were organized according to their respective impact indicators and transformed into the same unit for comparative assessment. The globally recognized SimaPro software and ecoinvent database were employed to evaluate the impacts (Marsmann and Schiburr, 1993; Pascual-González et al., 2016; Frischknecht and Rebitzer, 2005; Boustead, 1993). The cradle-to-gate environmental impacts were evaluated using the ReCiPe method under categories like global warming, ozone formation, ecotoxicity, water consumption, etc. The cradle-to-grave impacts were estimated by the Impact 2002 + method under four groups: resources, climate

Table 9.1 Life cycle inventory for LCA of the considered hydropower plants located in alpine and nonalpine regions.

Electricity, hydropower at the reservoir power plant, alpine region/RER U/AusSD U	**1 MJ**		
Inputs from nature	**Amount**		**Unit**
	Alpine	**Nonalpine**	
Transformation, from unknown	0.000023	0.00023	m^2
Transformation, to water bodies, artificial	0.0000228	0.000228	m^2
Transformation, to industrial area, built up	0.00000023	0.0000023	m^2
Occupation, water bodies, artificial	0.0035	0.035	m^2
Volume occupied, reservoir	0.15	0.15	m^3
Water, turbine use, unspecified natural origin	0.81	0.81	m^3
Energy, potential (in hydropower reservoir), converted	3.79	3.79	MJ
Energy, potential (in hydropower reservoir), converted	3.79	3.79	MJ
Inputs from technosphere			
Lubricating oil, at plant/RER U/AusSD U	0.000007	0.000007	kg
Reservoir hydropower plant, alpine region/RER/I U/AusSD U*Emissions to air*	3.90×10^{-13}		p
Dinitrogen monoxide	0.000000077		kg
Methane, biogenic	0.000014	2.86×10^{-4}	kg
Outputs to technosphere			
Disposal, used mineral oil, 10% water, to hazardous waste incineration/CH U/AusSD U	0.000007	0.000007	kg

change, ecosystem quality, and human health. In addition, the GHG emissions were assessed by the IPCC method over a 100-year time period. This method takes into account emissions like carbon dioxide, methane, nitrous oxide, etc. (Minx et al., 2017). The LCA methods (Impact 2002 + and IPCC) applied for this work are presented in Fig. 9.4. Finally, an uncertainty analysis was accomplished by the Eco-indicator 99 method to check the probability distribution of the considered cases for both plants.

The last step of LCA was the life cycle impact interpretation, where the effect outcomes were analyzed and compared with the main LCA

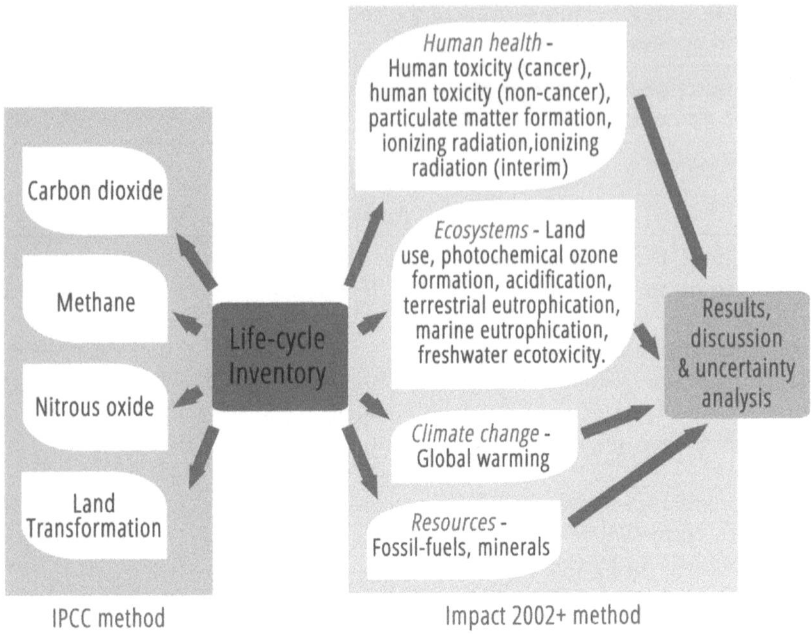

Figure 9.4 Life cycle assessment method.

goal of this research. At this stage, the main factors that account for both plants' and environmental hazards were recognized.

9.3 Results and discussion

9.3.1 Environmental profiles of the hydropower plants

The ReCiPe 2016 Midpoint (H) V1.00 method, the Impact 2002 + V2.14 method, and CML approaches were used to estimate the environmental performances of the alpine and nonalpine zone-based hydropower plants (Stavropoulos et al., 2016). The ReCiPe method follows the cradle-to-gate approach, which estimates and compares the environmental effects of the plants from raw material extraction to electricity production. On the other hand, the cradle-to-grave approach is used to obtain impacts in the Impact 2002 + method. The overall input—output comparison was obtained by the Raw material flow (RMF) method. Fig. 9.5 shows the RMF approach with natural inputs, environmental (air, water, soil) outputs, and solid waste.

As depicted, almost all of the inputs and outputs were alike, except for the outputs to air; plants in nonalpine areas (red bars) generate higher air emissions compared to the alpine-based plants (blue bars).

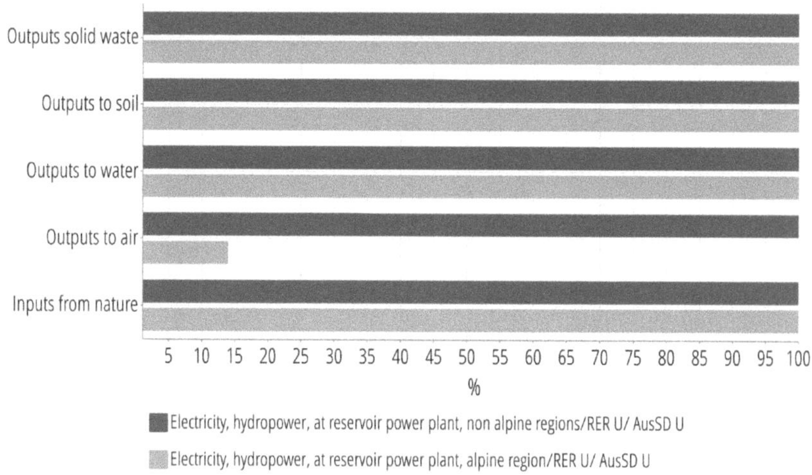

Figure 9.5 Comparative environmental impacts of the hydropower plants of alpine and nonalpine regions by the Raw material flow method.

Table 9.2 Comparative effects of the considered plants by ReCiPe method.

Impact category		Electricity, hydropower, at reservoir power plant, alpine region/RER U/AusSD U	Electricity, hydropower, at reservoir power plant, non-alpine regions/RER U/AusSD U	Unit
Global warming	Human health	1.7838999×10^{-9}	4.1655307×10^{-9}	DALY
	Terrestrial ecosystems	$5.3818149 \times 10^{-12}$	$1.2556865 \times 10^{-11}$	species.yr
	Freshwater ecosystems	$1.4702874 \times 10^{-16}$	$3.4298551 \times 10^{-16}$	species.yr
Ozone formation	Human health	$4.8714449 \times 10^{-12}$	$4.8714448 \times 10^{-12}$	DALY
	Terrestrial ecosystems	$7.0163971 \times 10^{-13}$	7.016397×10^{-13}	species.yr
Ecotoxicity	Terrestrial	$5.5926631 \times 10^{-14}$	5.592663×10^{-14}	species.yr
	Freshwater	$1.4610729 \times 10^{-15}$	$1.4610729 \times 10^{-15}$	species.yr
	Marine	5.143076×10^{-16}	5.143076×10^{-16}	species.yr
Water consumption	Human health	5.0869534×10^{-7}	5.0041953×10^{-6}	DALY
	Terrestrial ecosystem	3.0934176×10^{-9}	3.0430917×10^{-8}	species.yr
	Aquatic ecosystems	$1.3840179 \times 10^{-13}$	$1.3615018 \times 10^{-12}$	species.yr

Impact category	Electricity, hydropower, at reservoir power plant, alpine region/RER U/AusSD U	Electricity, hydropower, at reservoir power plant, nonalpine regions/RER U/AusSD U	Unit
Others			
Terrestrial acidification	$1.1623602 \times 10^{-12}$	$1.1623602 \times 10^{-12}$	species.yr
Freshwater eutrophication	3.762101×10^{-14}	3.762101×10^{-14}	species.yr
Stratospheric ozone depletion	$3.6271486 \times 10^{-13}$	$2.3780375 \times 10^{-13}$	DALY
Ionizing radiation	$1.0769103 \times 10^{-13}$	$1.0769103 \times 10^{-13}$	DALY
Fine particulate matter formation	1.2586791×10^{-9}	1.258679×10^{-9}	DALY
Human carcinogenic toxicity	$5.2830406 \times 10^{-10}$	$5.2830405 \times 10^{-10}$	DALY
Human noncarcinogenic toxicity	2.499131×10^{-11}	$2.4991309 \times 10^{-11}$	DALY
Land use	$3.7055138 \times 10^{-13}$	$3.7055138 \times 10^{-13}$	species.yr

The relative effects of the alpine- and nonalpine-based hydropower plants found on various factors attained by the ReCiPe method were grouped in Table 9.2. The values show that human health is the most

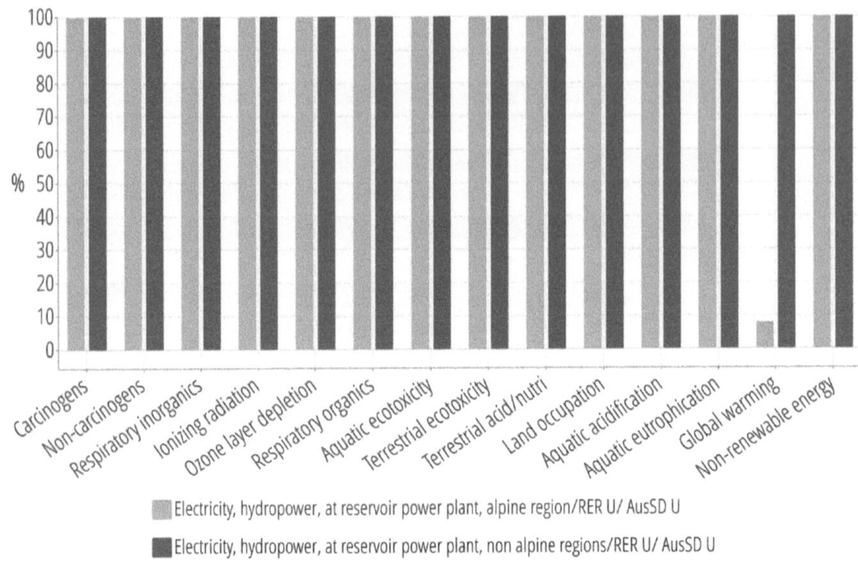

Figure 9.6 Comparison of overall environmental impacts of the considered plants by the Impact 2002 + method.

likely to be affected by the emissions from global warming. Additionally, hydropower plants in nonalpine areas have a greater share of hazardous global warming emissions affecting human health than plants in alpine areas. The effects of ozone formation-based impacts under the group of human health and terrestrial ecosystems have identical values for both considered plants. Environmental impacts in terms of marine, freshwater, and terrestrial ecotoxicity were also similar for both hydropower plants. Considering the water consumption impact, the tabulated value presents that human health was the most affected compared to aquatic and terrestrial ecosystems. In addition, alpine-zone-based hydropower plants provided better environmental profiles than nonalpine plants in this group. Environmental impacts of the ReCiPe method from other miscellaneous groups such as land use, human carcinogenic and noncarcinogenic toxicity, fine particulate matter formation, ionizing radiation, stratospheric ozone depletion, freshwater eutrophication, and terrestrial acidification are also presented in Table 9.2. As it was observed from the numerical values, the impacts were similar for both the considered plants. The highest impact category is the fine-particulate-matter formation and the human carcinogenic toxicity is in the second highest position in the same category.

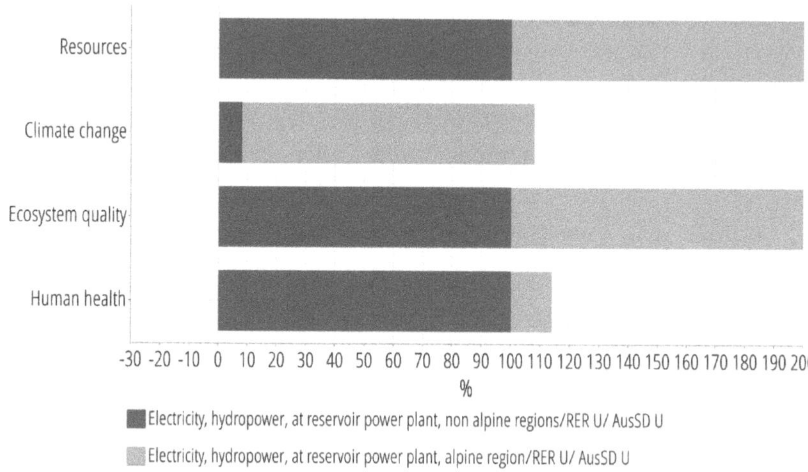

Figure 9.7 Endpoint damage assessment comparison of the plants using the Impact 2002 + approach.

The Impact 2002 + method was used to assess the environmental impacts of the alpine and nonalpine hydropower plants by the endpoint approach, as presented in Fig. 9.6. As depicted, 14 different impact parameters, such as carcinogens, noncarcinogens, respiratory inorganics, ionizing radiation, ozone layer depletion, respiratory organics, aquatic ecotoxicity, terrestrial ecotoxicity, terrestrial acid, land occupation, aquatic acidification, aquatic eutrophication, global warming, and non-renewable energy were evaluated by this method. It can be observed that apart from global warming, the other indicators provided a similar quantity. Alpine-region-based hydropower plants provided better impacts due to less GHG emissions than nonalpine region-based plants. Nevertheless, the 14 categories mentioned above assessed by the Impact 2002 + approach were categorized into four major types of effects: resources, climate change, ecosystem quality, and human health, as presented in Fig. 9.7. As depicted, the hydropower plants in nonalpine regions are more accountable for climate change compared to the hydropower plants in alpine regions. Nevertheless, the remaining three indicators are similar for both power plants.

The comparative impact assessment for both plants was examined with the use of the CML method considering 10 categories: abiotic depletion, global warming, ozone layer depletion, human toxicity, freshwater aquatic ecotoxicity, marine aquatic ecotoxicity, terrestrial ecotoxicity, photochemical oxidation, acidification, and eutrophication as depicted in Fig. 9.7. It can be observed that hydropower plants of nonalpine regions have almost double the impacts of global warming compared to that of alpine regions. Also, it can be depicted that hydropower plants in nonalpine regions are much more responsible for causing photochemical oxidation (secondary air pollution) than alpine regions. All the remaining indicators show almost the same value for both plants.

9.3.2 Metal- and gas-based emission evaluation

The Ecopoints 97 (CH) V2.07 method was applied for both of the cases to evaluate the metal- and gas-based emissions. The Ecopoints 97 method is recognized as a structured means for assessing any product's metal and hazardous gas emissions (Pesso, 1993). Twenty-one different indicators were evaluated by this method. These are carbon dioxide (CO_2), nitrous oxide (NO_x), sulfur oxide (SO_x), ammonia (NH_3), nitrogen (N), copper (Cu), cadmium (Cd), Nickel (Ni), mercury (Hg), zinc (Zn), lead (Pb), nitrate, pesticide, waste, phosphorous (P), dissolved organic carbon (COD), chromium (Cr), non-methane volatile organic compounds, dust PM10, LMRD, HRAD, and energy with a set highest rate of 100%. The total outcome from the Ecopoints 97 method is illustrated in Fig. 9.8. As observed from the figure, alpine zone-based hydropower plants generate the lowest quantity of CO_2 emissions against the hydropower plants of nonalpine regions, leaving the remaining indicators identical for both plants.

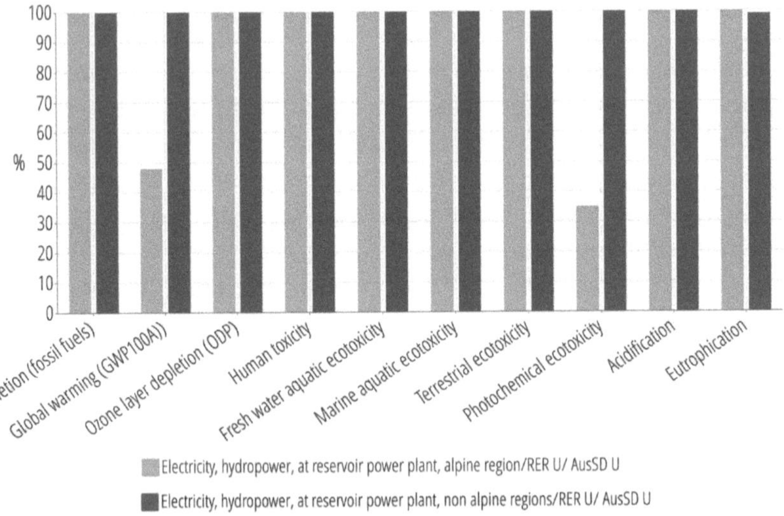

Figure 9.8 Comparative environmental impacts of the hydropower plants of alpine and nonalpine regions by the CML approach.

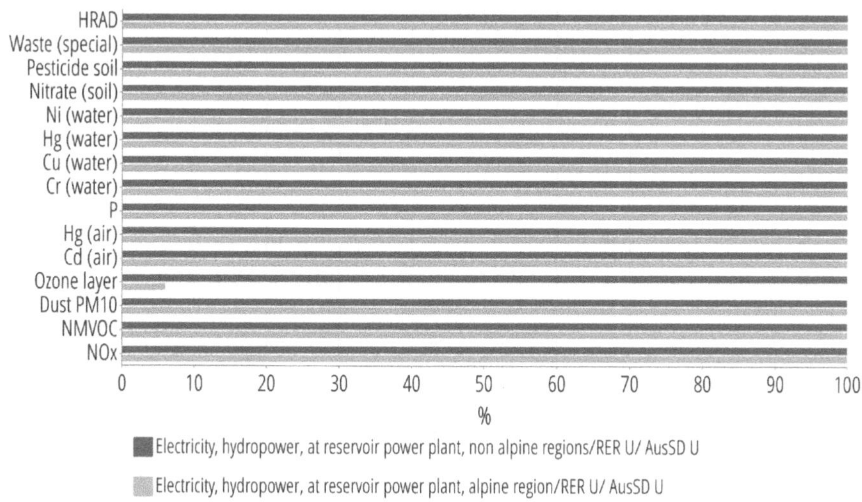

Figure 9.9 Comparative metal and gas-based emissions by the considered plants using the Ecopoints 97 method.

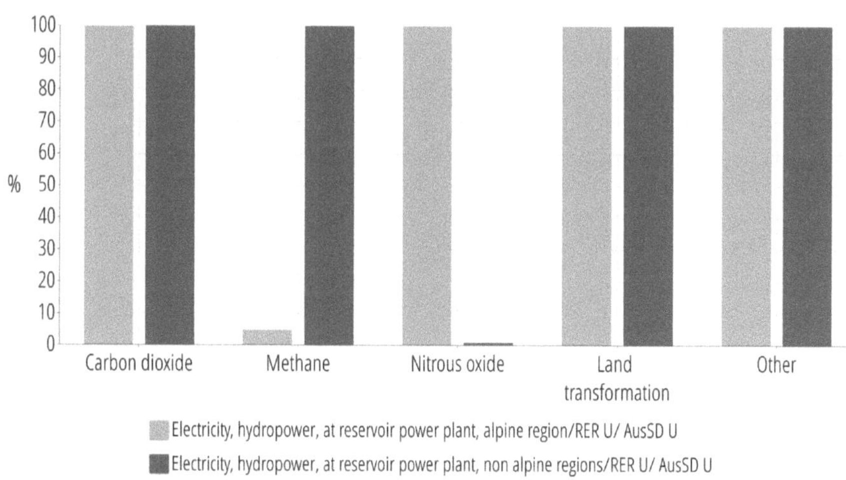

Figure 9.10 Comparison of the greenhouse gas emissions by the plants, following the IPCC approach.

9.3.3 Greenhouse-gas emission estimation

The comparative analysis of GHG emissions by hydropower plants of alpine and nonalpine areas of Europe from the IPCC method is presented in Fig. 9.9, where the time span is 100 years. The IPCC technique is considered a well-structured system to assess the characteristics of climate change (Minx et al., 2017). It can be seen that a similar amount of GHG emissions occur for both the selected plants under the group of carbon dioxide and land transformation. However, nonalpine zone-based hydropower plants emit more methane, while nitrous oxide is emitted to a great extent from the alpine zone-based power plants.

9.3.4 Uncertainty analysis

Figs. 9.10 and 9.11 represent the uncertainty analysis for the hydropower plants in alpine and nonalpine zones, respectively. The Eco-indicator 99

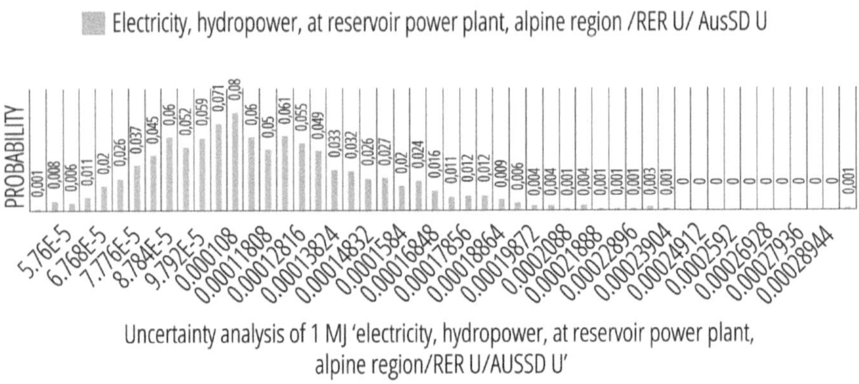

Figure 9.11 Probability distribution for the single-score impact category of hydropower plants of alpine zones.

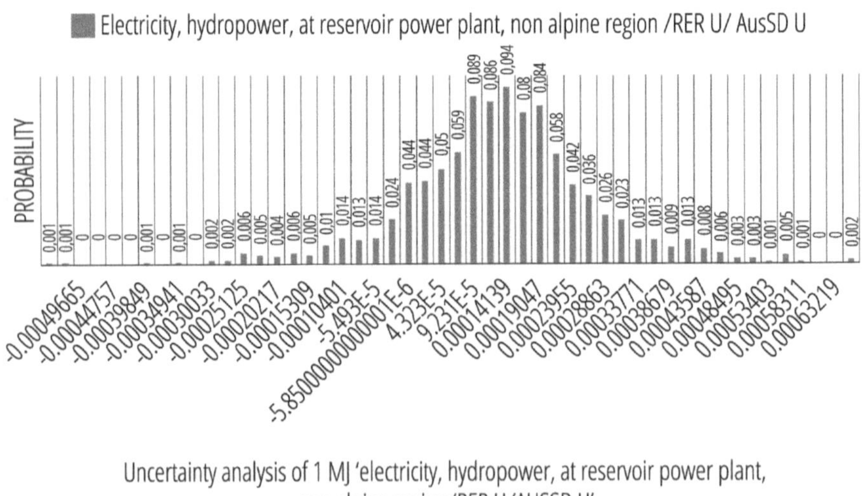

Figure 9.12 Probability distribution for the single-score impact category of hydropower plants of nonalpine zones.

method was employed in assessing the uncertainty distributions. The possibilities increase with an increase in the bar's height and the maximum height bars with the maximum height present the highest possibilities. The small bars for about 70% of the total numbers indicate that hydropower plants of both regions are environmentally superior (Fig. 9.12).

9.4 Conclusion

This paper presents a comparative assessment of the environmental impacts of alpine and nonalpine region-based hydropower plants in Europe. The potency of this LCA analysis was validated by some parameters: (1) development of an upright all-inclusive LCI for both the selected plants, (2) impact assessment through a number of methods, and (3) performing an uncertainty analysis from both perspectives to evaluate the relative efficiencies. This research task was eccentric since it was estimated the ecological effects using different systematic methodologies, which have not been performed before. Also, the research techniques were considered both the midpoint and the endpoint approaches in assessing the effects from both installations, utilizing the Impact 2002 + method. The results were showed that alpine zone-based hydropower plants provide a superior environmental profile compared to nonalpine zones. In addition, the assessment of the GHG emissions was provided that hydropower plants in alpine zones emit about 10 times lower methane gas in comparison to nonalpine zone-based power plants. All in all, it is proposed by this work that in order to elevate the generation of clean hydropower in the future, further research is needed to be more focused on detecting the most dangerous materials during plant installation and replacing them with matching ecologically superior options.

References

Armaroli, N., Balzani, V., 2011. Towards an electricity-powered world. Energy & Environmental Science 4 (9), 3193–3222.

Atilgan, B., Azapagic, A., 2015. Life cycle environmental impacts of electricity from fossil fuels in Turkey. Journal of Cleaner Production 106, 555–564.

Botelho, A., Ferreira, P., Lima, F., Pinto, L.M.C., Sousa, S., 2017. Assessment of the environmental impacts associated with hydropower. Renewable and Sustainable Energy Reviews 70, 896–904.

Boustead, I., 1993. General principles for life cycle assessment databases. Journal of Cleaner Production 1 (3), 167–172.

Capellán-Pérez, I., Arto, I., Polanco-Martinez, J.M., González-Eguino, M., Neumann, M. B., 2016. Likelihood of climate change pathways under uncertainty on fossil fuel resource availability. Energy & Environmental Science 9 (8), 2482–2496.
Curran, M.A., 2013. Life cycle assessment: a review of the methodology and its application to sustainability. Current Opinion in Chemical Engineering 2 (3), 273–277.
Das, D., Srinivasan, R., Sharfuddin, A., 2011. Fossil fuel consumption, carbon emissions and temperature variation in India. Energy & Environment 22 (6), 695–709.
de Haes, H.A.U., 1993. Methodology and LCA applications. Journal of Cleaner Production 1 (3–4), 205–206.
de Larderel, J.A., 1993. Environmental life cycle assessment and its applications. Journal of Cleaner Production 1 (3), 130.
de Miranda Ribeiro, F., Da Silva, G.A., 2010. Life-cycle inventory for hydroelectric generation: a Brazilian case study. Journal of Cleaner Production 18 (1), 44–54.
Espinosa, N., Hösel, M., Jørgensen, M., Krebs, F.C., 2014. Large scale deployment of polymer solar cells on land, on sea and in the air. Energy & Environmental Science 7 (3), 855–866.
Farjana, S.H., Huda, N., Mahmud, M.P., Saidur, R., 2018. Solar process heat in industrial systems—a global review. Renewable and Sustainable Energy Reviews 82, 2270–2286.
Finkbeiner, M., Inaba, A., Tan, R., Christiansen, K., Klüppel, H.J., 2006. The new international standards for life cycle assessment: ISO 14040 and ISO 14044. The International Journal of Life Cycle Assessment 11 (2), 80–85.
Frischknecht, R., Rebitzer, G., 2005. The ecoinvent database system: a comprehensive web-based LCA database. Journal of Cleaner Production 13 (13–14), 1337–1343.
Gong, J., Darling, S.B., You, F., 2015. Perovskite photovoltaics: life-cycle assessment of energy and environmental impacts. Energy & Environmental Science 8 (7), 1953–1968.
Heijungs, R., Guinée, B., 2012. An overview of the life cycle assessment. Life Cycle Assessment Handbook: A Guide for Environmentally Sustainable Products. Wiley, pp. 15–44.
Keoleian, G.A., 1993. The application of life cycle assessment to design. Journal of Cleaner Production 1 (3–4), 143–149.
Lazzaro, G., Botter, G., 2015. Run-of-river power plants in Alpine regions: whither optimal capacity? Water Resources Research 51 (7), 5658–5676.
Lewandowska, A., Matuszak-Flejszman, A., Joachimiak, K., Ciroth, A., 2011. Environmental life cycle assessment (LCA) as a tool for identification and assessment of environmental aspects in environmental management systems (EMS) Part 2—case Studies. The International Journal of Life Cycle Assessment 16 (3), 247–257.
Li, Z., Du, H., Xiao, Y., Guo, J., 2017. Carbon footprints of two large hydro-projects in China: life-cycle assessment according to ISO/TS 14067. Renewable Energy 114, 534–546.
Mahmud, M.P., Farjana, S.H., 2012. Wind power technology schemes as renewable energy in bangladesh. International Journal of Engineering and Advanced Technology (IJEAT) 1, 315–319.
Mahmud, M.P., Lee, J., Kim, G., Lim, H., Choi, K.B., 2016. Improving the surface charge density of a contact-separation-based triboelectric nanogenerator by modifying the surface morphology. Microelectronic Engineering 159, 102–107.
Marsmann, M., Schiburr, A., 1993. Databases, software and LCA applications. Journal of Cleaner Production 1 (3–4), 206–207.
Minx, J.C., Callaghan, M., Lamb, W.F., Garard, J., Edenhofer, O., 2017. Learning about climate change solutions in the IPCC and beyond. Environmental Science & Policy 77, 252–259.

Parvez Mahmud, M.A., Huda, N., Farjana, S.H., Asadnia, M., Lang, C., 2018. Recent advances in nanogenerator-driven self-powered implantable biomedical devices. Advanced Energy Materials 8 (2), 1701210.

Parvez Mahmud, M.A., Farjana, S.H., 2012. Design and construction of refrigerant charge level detecting device in hvac/r system with microcontroller. International Journal of Engineering and Advanced Technology 1 (5), 2249–8958.

Pascual-González, J., Guillén-Gosálbez, G., Mateo-Sanz, J.M., Jiménez-Esteller, L., 2016. Statistical analysis of the ecoinvent database to uncover relationships between life cycle impact assessment metrics. Journal of Cleaner Production 112, 359–368.

Pesso, C., 1993. Life cycle methods and applications: issues and perspectives. Journal of Cleaner Production 1 (3–4), 139–142.

Pryshlakivsky, J., Searcy, C., 2013. Fifteen years of ISO 14040: a review. Journal of Cleaner Production 57, 115–123.

Périlhon, C., Alkadee, D., Descombes, G., Lacour, S., 2012. Life cycle assessment applied to electricity generation from renewable biomass. Energy Procedia 18, 165–176.

Santoyo-Castelazo, E., Gujba, H., Azapagic, A., 2011. Life cycle assessment of electricity generation in Mexico. Energy 36 (3), 1488–1499.

Stavropoulos, P., Giannoulis, C., Papacharalampopoulos, A., Foteinopoulos, P., Chryssolouris, G., 2016. Life cycle analysis: comparison between different methods and optimization challenges. Procedia CIRP 41, 626–631.

Turconi, R., Boldrin, A., Astrup, T., 2013. Life cycle assessment (LCA) of electricity generation technologies: overview, comparability and limitations. Renewable and Sustainable Energy Reviews 28, 555–565.

Uddin, M.S., Kumar, S., 2014. Energy, emissions and environmental impact analysis of wind turbine using life cycle assessment technique. Journal of Cleaner Production 69, 153–164.

Wagner, T., Themeßl, M., Schüppel, A., Gobiet, A., Stigler, H., Birk, S., 2017. Impacts of climate change on stream flow and hydro power generation in the Alpine region. Environmental Earth Sciences 76 (1), 4.

Ward, D.J., Inderwildi, O.R., 2013. Global and local impacts of UK renewable energy policy. Energy & Environmental Science 6 (1), 18–24.

Židonienė, S., Kruopienė, J., 2015. Life cycle assessment in environmental impact assessments of industrial projects: towards the improvement. Journal of Cleaner Production 106, 533–540.

Zou, C., Zhao, Q., Zhang, G., Xiong, B., 2016. Energy revolution: from a fossil energy era to a new energy era. Natural Gas Industry B 3 (1), 1–11.

CHAPTER 10

A stakeholder impact analysis of the production of the energy vector hydrogen

Holger Schlör[1] and Sandra Venghaus[1,2]
[1]Forschungszentrum Jülich, Jülich, Germany
[2]School of Business and Economics, RWTH Aachen University, Aachen, Germany

Contents

10.1 Introduction

The United Nations (UN) has called for significant changes to the current unsustainable consumption and production patterns (United Nations, 2012) to mitigate climate change and enable sustainable development (United Nations, 2015). Since 2011, the German government has supported the UN (UNEP, 2011) and European Union (Newbery, 2016) efforts with the transformation of the German economy towards a green economy (Federal Ministry of Education & Research (BMBF), 2014) with the objective to protect the well-being of current and future generations (German Federal Government, 2010). Currently, the transformation process of the German energy sector (Energiewende) (Hake et al., 2015; German Federal Ministry of Economics & Technology BMWI, 2012) is significantly shaped by the impending nuclear and coal phase-out[1] (Bundesministerium für Wirtschaft

[1] https://www.bmwi.de/Redaktion/DE/Artikel/Service/Gesetzesvorhaben/kohleausstiegsgesetz.html

Environmental Assessment of Renewable Energy Conversion Technologies
DOI: https://doi.org/10.1016/B978-0-12-817111-0.00011-5

und Energie BMWI, BMWI, 2014; Bundesministerium für Wirtschaft und Energie BMWI, 2016). Green energy technologies are considered central instruments in this transformation process (German Federal Government, 2017). In the view of the German government, the hydrogen economy (Campbell, 2014; Gregory, 1973) provides an opportunity to replace the current fossil-fuel-based economy (Abe et al., 2019; Muradov & Veziroğlu, 2005). According to Navlani-García et al. (2018), it is essential to develop renewable and clean energy alternatives to meet the global energy demand while mitigating environmental impacts, proposing hydrogen to be a promising option. In the past, hydrogen has often been praised for its potential to replace the central role of fossil energy in a 21st century economy (Rifkin, 2003). Already in 1974 and 1976, at the Miami Beach hydrogen conferences,[2] NASA proposed that the hydrogen economy will come (Hoffmann, 2012), assuming the hydrogen economy to be a model for energy security and environmental protection (Hoffmann, 2012).

However, hydrogen in itself is not an energy source but rather an energy carrier that must be produced (Hoffmann, 2012) and which is defined as an energy vector, which "allows to transfer, in space and time, a quantity of energy", thus enabling the transport of the produced renewable energy across countries and regions, as well as storing it over time (Orecchini, 2006).

More specifically, the following characteristics make hydrogen a frequently proposed key cornerstone in the transformation of energy systems (Bleischwitz et al., 2010; Ren et al., 2017; Navlani-García et al., 2018):

- It is an energy vector (Orecchini, 2006) that can be produced from many renewable energy sources, including biomass, solar, or wind, and it can be stored (Zhang et al., 2018; Cano et al., 2018) (intermittent renewable energy production (Bleischwitz et al., 2010; Hoffmann, 2012).
- It can be synthesized using different processes and even when burnt, its combustion product is mostly water (Gregory, 1973).
- It can flexibly be integrated into different transformation pathways and transition strategies (Bleischwitz et al., 2010).

Hydrogen can be used (Krajačić et al., 2008; Orecchini, 2006) in road transport as currently demonstrated by the Toyota Mirai,[3] the Hyundai

[2] https://www.bmbf.de/de/nationale-wasserstoffstrategie-9916.html

[3] Hydrogen Economy Miami Energy Conference; March 18, 1974 - March 20, 1974; Miami Beach, FL (https://ntrs.nasa.gov/search.jsp?R=19750060704) and 1st World Hydrogen Energy Conference; March 01, 1976 - March 03, 1976; Miami Beach, FL (https://ntrs.nasa.gov/search.jsp?R=19770050566).

ix35 Fuel Cell Electric Vehicle[4] (FCEV), the Hyundai Nexo,[5] the Mercedes GLC F-CELL[6] (electric vehicle with fuel cell), and hydrogen buses (O'Garra et al., 2007).

However, it is important to note that hydrogen and its use in fuel cells can only be ecologically sustainable if hydrogen is produced from renewable energy sources (Bleischwitz et al., 2010). In addition, only under these conditions will it meet the social standards of the UN as formalized in the UN sustainable development goals (Nations, 2015; Dou et al., 2017).

According to the International Energy Agency (Drennen, 2013), it is essential to analyze the hydrogen economy using a systems analysis approach in order to capture the different characteristics and their socio-economic-ecological effects within the involved sectors, specifically the transport and energy sectors. The used system approach rests on the definition of Donella Meadows, who describes a system as "an interconnected set of elements that is coherently organized in a way that achieves something" (Meadows, 2008). According to this definition, a system consists of three parts: elements, interconnections, and a function or purpose. Applied to the hydrogen economy, the stakeholders (workers, local community, and society) represent the elements of the system, the physical energy flows represent the interconnections, and reducing CO_2 emissions represents the purpose. Meadows further adds to this definition that the flow of information is the precondition for operating the interconnections in systems (Meadows, 2008).

Indicators are needed to systematically capture the complex interrelationships of the hydrogen system to make explicit the flow of information within the system and the surrounding energy system (Penner, 2006), which are embedded in the socio-economic-ecological system (Venghaus, 2011). The indicators serve as the fundament for identifying the structure of the hydrogen-economy-system, consisting of the elements of the system, its interconnections, as well as the influencing factors on system behavior (Marbán & Valdés-Solís, 2007; Rifkin, 2003; Moliner et al., 2016; Moreno-Benito et al., 2017; Dagdougui et al., 2018).

4 https://h2.live/wasserstoffautos/hyundai-ix35
5 https://h2.live/fahren
6 https://www.daimler.com/innovation/case/electric/brennstoffzelle-wasserstoff-elektromobilitaet.html

This analysis was concentrated on the hydrogen system elements: the stakeholders and indicators to describe the elements of the system. For the specific analysis, the hydrogen process chain—the core of the hydrogen economy—was analyzed with respect to its role in the social context using a Social Hotspot Database (SHDB)-based model. This approach enables measuring the social impact of the hydrogen production chains by making explicit the social risks. It was analyzed whether the social foundation of hydrogen production is stable and what social consequences are expected based on certain production and value chains.

10.2 Methodological framework and background

In 2009, Nicolas Sarkozy stressed two main arguments in his foreword to the Stiglitz-Sen-Fitoussi-Commission report (Stiglitz et al., 2009), summarizing the necessity of developing new measuring methods for determining the quality of life in the context of sustainable development in a globalized world (Stiglitz et al., 2010).

- "We will not change our behavior unless we change the ways we measure our economic performance. [...] We must change the criteria governing our social organizations and our public policies (Stiglitz et al., 2010)" to avoid socio-economic and ecological crises.
- "Our statistics and accounts reflect our aspirations, the values that we assign things. They are inseparable from our vision of the world and the economy, of society, and our conception of human beings and our interrelations." Sarkozy encourages us to interpret our results not only as objective data because we always have to ask ourselves "about the purpose of what we are doing, what we are actually measuring, and what lessons we need to draw" (Stiglitz et al., 2010).

Building on the above argumentation, this research approach was developed to demonstrate that new bottom-up measuring approaches of our social assessment model enable us to assess the social effects of new energy vector technologies on the stakeholders' quality of life. We elucidate how their well-being is influenced by hydrogen production.

For our analysis, a careful description of the various elements and steps of hydrogen production in Germany, Austria, and Spain based on the analysis of Hake et al., Kuckshinrichs et al., and Koj et al. is used (Kuckshinrichs & Koj, 2018; Koj et al., 2015; Hake et al., 2017) for the social assessment model based on the current power mix of Germany, Austria, and Spain.

10.3 Data—social hotspot database

Only a few suitable social databases are currently available. For the analysis, the SHDB of New Earth (Norris et al., 2014; Benoit-Norris et al., 2012; van Haaster et al., 2017) was chosen, which assigns social risk points to each of the production steps (New Earth/ Social Hotspots Database Project, 2014).

The SHDB enables

1. the modeling of production systems, and
2. the initial assessment of potential social impacts in social risk points.

It thus rests on a "background net that captures and represents the full supply chain and its potential impacts" (Benoît Norris, 2014).

The SHDB originated from work on the United Nations Environment Programme (UNEP)/Society of Environmental Toxicology and Chemistry (SETAC) Social Life Cycle Assessment (LCA) Guidelines, 2009 (Sonnemann and Valdivia, 2014). It was developed by the NGO New Earth (New Earth/ Social Hotspots Database Project, 2014; Norris et al., 2014).

The SHDB includes:

- A Global Input/Output model[7] derived from the Global Trade Analysis Project Version 7 (GTAP)[8] of 2008. A 113-region static input/output model containing data pertaining to 57 economic sectors (GTAP Global Trade Analysis Project, 2008).
- A worker hours model, which is constructed from annual wage payments and wage rates by country and sector (Benoît-Norris et al., 2013): "The labor intensity data were developed by converting GTAP data on wage payments into estimates of worker hours [. . .] for each sector in each GTAP country/region by compiling and using wage rate data [. . .] by sector and region (Benoît Norris, 2014)."
- Social theme tables, which cover five social impact categories—labor rights, health and safety, human rights, governance, and community infrastructure—are capturing social risks: "The social risk data was collected from over 200 reputable publicly available sources over a period of 4 years. The data tables are updated every 2 years with the newest information. Several criteria are used to select the data to be incorporated into the SHDB (Benoît Norris, 2014)."

[7] The GTAP we used contains the 2004 database, which was published 2008 GTAP (Global Trade Analysis Project) 2008. GTAP 7 database. *In:* Purdue University (2008) (ed.) *GTAP*. Purdue University.

[8] http://www.gtap.agecon.purdue.edu/databases/

- The SHDB is a global database (van Haaster et al., 2017) that provides for identifying "hotspots" of "production activities in the supply chain that may be at risk for social issues (Benoit-Norris et al., 2012)." The SHDB contains social quantitative and qualitative information, weighted according to risk (low, medium, high, and very high) for 227 countries and 57 sectors (Norris et al., 2014). "In total, over 400 publicly available data sources have been reviewed and over 200 incorporated into the database."[9]
- The SHDB enables social conditions to be examined in these countries based on 100 indicators. This database was built on data from the period between 2004 and 2012.[10]

The UNEP/SETAC guidelines (UNEP/SETAC Society of Environmental Toxicology and Chemistry, 2009) and the SHDB (Ciroth and Franze, 2011) allow using both qualitative and quantitative data sets (UNEP/SETAC Life Cycle Initiative, 2009; Ciroth and Franze, 2011). Specifically, they were analyzed for those five major social impact categories (labor rights, health and safety, human rights, governance, community and infrastructure) of the SHDB (United Nations Development Programme (UNDP), 1994), which affect the stakeholders. Each category [. . .] contains a range of specific indicators within multiple social themes. Either a single indicator or several related indicators are used to characterize the risk of a particular social issue (Benoît-Norris et al., 2013). In the following, the LCA data set used for the stakeholder analysis will be described.

10.4 Hydrogen production simplified process chain—system boundaries of the hydrogen process chain

For the stakeholder analysis, the LCA data set of Hake et al. (2017) and Koj et al. (2017) serve as the basis to describe the technical hydrogen process chain. They analyzed the pressurized alkaline water electrolysis (AEL) systems operating in Austria, Germany, and Spain using a classical four stages LCA (goal and scope definition, inventory analysis, impact assessment, and interpretation) according to the LCA ISO norms (ISO 14040, 14044) to identify the individual process chains of the three production sites (Hake et al., 2017; Koj et al., 2017). This LCA served as the fundament for this stakeholder analysis. The technical system boundary of the LCA developed by the authors for the three AEL systems is presented in Fig. 10.1.

[9] GreenDelta, Social Hot Spots Database in openLCA, Quick explanation.
[10] GreenDelta, Social Hot Spots Database in openLCA, Quick explanation.

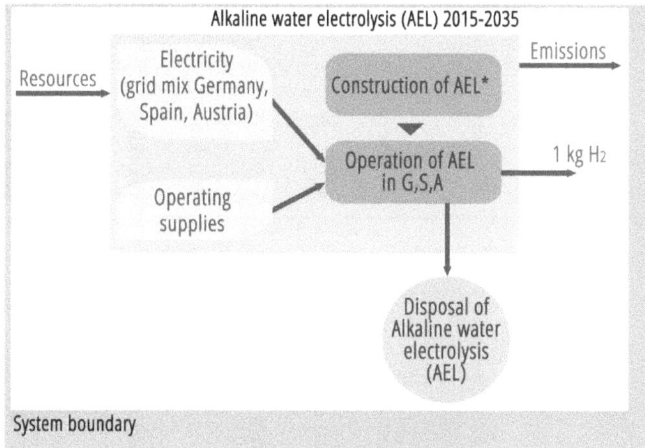

Figure 10.1 Technical system boundary of the 6 MW water electrolysis system.

It is assumed that the life span of the electrolyzer is 20 years, while individual components (cell stack) may be substituted during this period (Hake et al., 2017). Furthermore, for consistency reasons, the following assumptions were resumed from the original study:

> *"The production of 1 kg of H_2 (33 bar, 40°C, 99.8% purity) is used as a consistent reference unit (functional unit) for all cases and all base methods. Hydrogen compression, storage, and further use are not considered within the system boundaries. Due to a lack of knowledge about recycling and disposal of the membranes, it was not possible to consider this aspect. Inventory data as the framework of all three assessments are mainly taken from 'ELYGRID' (German Federal Government, 2010). This includes the amount of electricity and other operating supplies such as deionized water, potassium hydroxide (KOH) and steam as well as systems engineering of AEL (Hake et al., 2017)."*

Fig. 10.1 describes the technical system boundaries for this stakeholder analysis of hydrogen production in Austria, Germany, and Spain.

10.5 Results—social risks of the stakeholders

For the stakeholder asssessment model (SAM) the SHDB (New Earth/ Social Hotspots Database Project, 2014) is used to calculate the total social hotspot risk points of hydrogen production in Germany, Austria, and Spain. The model analyzes the social risks based on 23 selected stakeholder assessment indicators from the SHDB, as Table 10.1 shows, to assess the living situation of the stakeholders (Benoît-Norris et al., 2013).

Table 10.1 Stakeholder assessment indicators based on SHDB.

	SHDB impact categories	Social themes	Issue—social indicators	Stakeholders
1	Labor rights + decent work	Wage assessment	Risk of average wage being lower than country's minimum wage	Workers
2	Labor rights + decent work	Forced labor	Risk of forced labor by sector	Workers
3	Labor rights + decent work	Migrant workers	Risk that migrant workers are treated unfairly	Workers
4	Labor rights + decent work	Freedom of association, collective bargaining, and right to strike	Risk that a country lacks or does not enforce the right to strike	Workers
5	Labor rights + decent work	Poverty	Risk of wages being under $2 per day	Workers
6	Labor rights + decent work	Unemployment	Risk of unemployment in country	Workers
7	Labor rights + decent work	Labor laws and conventions	Risk that country does not provide adequate labor laws	Workers
8	Labor rights + decent work	Child labor	Risk of total child labor in country	Workers
9	Labor rights + decent work	Working time	Risk of excessive working time by sector	Workers
10	Health and safety	Occupational injuries and deaths	Risk of fatal injuries by sector	Workers
11	Health and safety	Occupational toxics and hazards	Risk of workplace noise exposure, both genders	Workers
12	Human rights	Human health—noncommunicable diseases and their health risks	Risk of low life expectancy	Society

13	Human rights	Human health—communicable diseases	Risk of mortality from communicable disease	Local communities
14	Human rights	Gender equity	Overall risk of gender inequality in the country	Society
15	Human rights	High conflict zones	Characterization of Heidelberg Barometer1	Society
16	Human rights	Indigenous rights	Characterization of indigenous population	Society
17	Governance	Corruption	Overall risk of corruption considering all indicators	Society
18	Governance	Legal system	Risk of fragility of the legal system considering all indicators	Society
19	Community and infrastructure	Access to hospital beds	Risk that there are too few hospital beds to support population	Local communities
20	Community and infrastructure	Access to improved drinking water	Risk of no access to an improved source of drinking water	Local communities
21	Community and infrastructure	Smallholder vs commercial labor	Characterization of farm labor	Workers
22	Community and infrastructure	Access to improved sanitation	Risk of no access to an Improved source of sanitation	Local communities
23	Community and infrastructure	Children out of school	Risk that children do not attend school	Local communities

Source: Adapted from Benoît-Norris, C., Norris, G.A., Aulisio Cavan, D. (2013). Social hotspot database. Supporting documentation update: 2013. York Beach, Maine, New Earth. Maine: SHDB.

The stakeholder assessment model consists of five social impact categories (labor rights and decent work, health and safety, human rights, governance, and community and infrastructure), which contain twenty three social themes. The social themes are concretized by 23 social indicators to measure the social risks for the three stakeholder groups: workers, local community, and society (Norris et al., 2014; Benoît et al., 2007). The workers are described by twelve indicators, society by six indicators, and the local communities and their infrastructure by five indicators.

The indicators were weighted equally with one according to the weighting guidelines of the New Earth/ Social Hotspots Database Project (2014). Fig. 10.2 shows that the three stakeholders of the three process chains are affected differently by hydrogen production. For the social group of "workers," the social risk is highest in the case of the Austrian hydrogen production chain (31.4), followed by the Spanish (27.5), and the German (11.2) production chains, respectively.

For "local communities" involved in the hydrogen process chains, those related to the Austrian chain face the highest social risk (18.2), followed by the Spanish chain (9.6). In the case of the German process chain (7.7), the social risk is less than half of the Austrian chain.

"Society" as a whole faces a social risk of 12.1 risk points in the case of the Austrian process chain, thus nearly twice that of the German chain

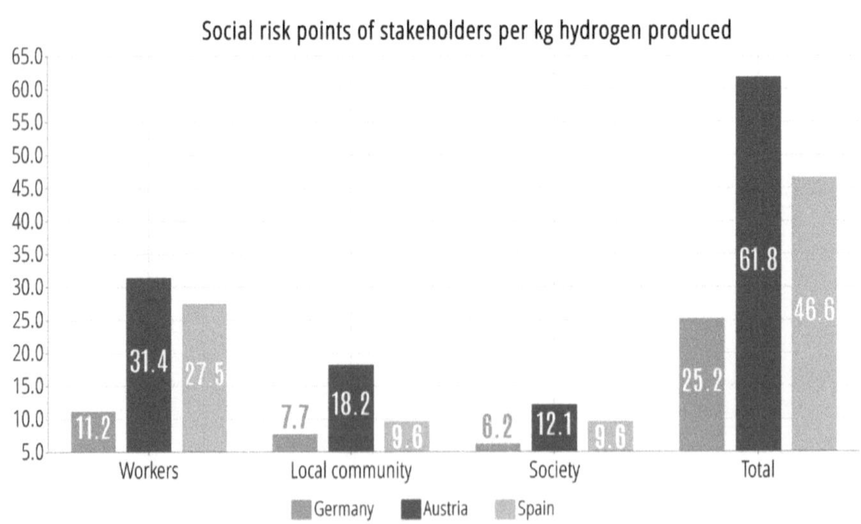

Figure 10.2 Social risk points of stakeholders per kilogram hydrogen produced.

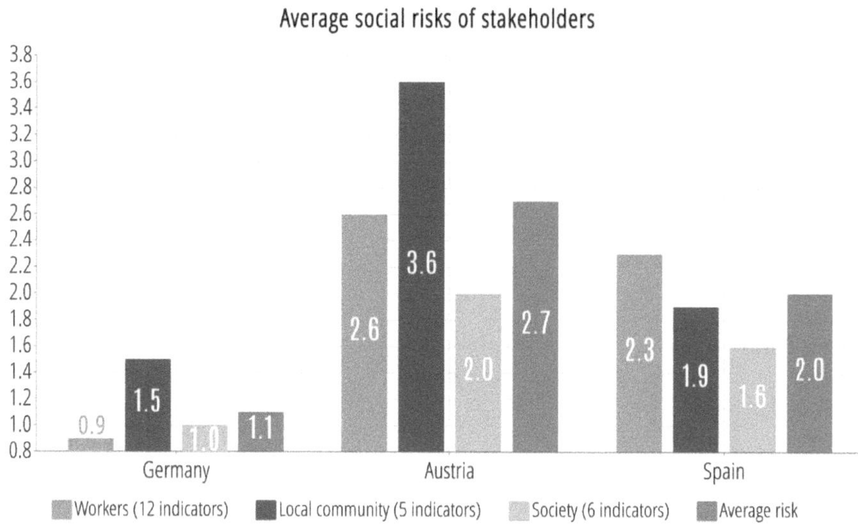

Figure 10.3 Average social risks of stakeholders.

with 6.2 risk points. The Spanish hydrogen process chain, in contrast, implies a social risk of 9.6 risk points per kilogram of hydrogen produced.

In summary, the total weighted social risk points for all stakeholder groups amount to 25.2 for the hydrogen production chain in Germany, 46.6 for the production chain in Spain, and 61.8 for the hydrogen production chain located in Austria.

The analysis of the average social risk for the various stakeholder groups (Fig. 10.3) detects an average social risk per stakeholder indicator of 2.7 in the case of the Austrian chain, average social risk of 2.0 points in the case of the Spanish chain, and average social risk of only 1.1 in the case of the German production chain.

When considering the average social risks per indicator for the different stakeholder groups (Fig. 10.3), a slightly different picture about the distribution of the risks is obtained. In this case, the "local communities" face the greatest risks (Austria: 3.6, Spain: 1.9, and Germany: 1.5), followed by the group of "workers" (Austria: 2.6, Spain: 2.3, and Germany: 0.9). "Society" overall faces a social risk of 2.0 risk points in Austria, 1.6 risk points in Spain, and 1.0 risk points in Germany.

The reason for this distribution across the three production sites results from differences in the imported social risks, which accordingly depend on the social conditions (i.e., the social risks) in the respective countries, from which intermediate products for hydrogen production are imported (Table 10.2).

Table 10.2 Affected stakeholders in countries causing social risks of hydrogen production estimation for the three countries—per kilogram of hydrogen production.

Countries	Social risk points in Austria*	Social risk points in Germany*	Social risk points in Spain*
Uzbekistan total	14.14	0.37	0.36
Workers	7.19	0.17	0.21
Local community	4.18	0.11	0.07
Society	2.78	0.09	0.07
China total	4.03	3.85	1.52
Workers	2.05	1.72	0.90
Local community	1.19	1.18	0.31
Society	0.79	0.95	0.31
Austria total	3.57	0.00	0.00
Workers	1.82		
Local community	1.06		
Society	0.70		
Russia total	2.91	0.96	1.16
Workers	1.48	0.43	0.68
Local community	0.86	0.29	0.24
Society	0.57	0.24	0.24
Algeria total	1.12	0.47	9.81
Workers	0.57	0.21	5.78
Local community	0.33	0.14	2.01
Society	0.22	0.12	2.02
Germany total	1.11	2.15	0.00
Workers	0.56	1.09	
Local community	0.33	0.64	
Society	0.22	0.42	
Spain total			3.59
Workers			2.11
Local community			0.73
Society			0.74
Total risk points (%)	5.8	8.5	7.7

*based on the 2013 data of the SHDB.

The social risks thus depend on the specific trade relations of the three hydrogen-producing countries. The following countries are involved in the hydrogen production process through trade relations: Uzbekistan, Egypt, China, Ukraine, Austria, Germany, Algeria, Russia, India,

Colombia, Angola, Romania, Mozambique, Equatorial Guinea, South Africa, Spain, Nigeria, Indonesia, Qatar, Kazakhstan, Nepal, U.K., and Turkey. Their stakeholders are affected differently by the production of the intermediate products for hydrogen production in the three countries, as the model estimations of Table 10.2 show for the stakeholders in the most strongly affected countries.

In the case of *Austria*, hydrogen production is connected to significant social risks imported from Uzbekistan (14.4 risk points), which is considerably higher than the social risks prevailing in China (4.03), Austria (3.57), Russia (2.91), and Egypt (2.57). The risks in India (1.94), Algeria (1.12), and Germany (1.11) are below these risks, whereas the risks in Ukraine, Mozambique, Kazakhstan, Indonesia, and Angola are significantly lower than in the other countries.

The case of *Germany* draws a different picture. The highest social risk imported with the intermediate products to Germany is from China (3.85), followed by the domestic German risks (2.15). India (2.04), Russia (0.96), Indonesia, Algeria, Mozambique, Uzbekistan, Kazakhstan, Colombia, Angola, Nepal, South Africa, and Turkey are causing even lower social risks in the case of German hydrogen production.

The hydrogen production in *Spain* is connected to social risks, mainly in Africa (Table 10.2). Specifically, the highest social risk is located in Algeria (9.81), followed by Spain (3.59), Angola (3.52), China (1.52), South Africa (1.28), Russia (1.16), and Indonesia (1.06). The social risks in India, Equatorial Guinea, Ukraine, Nigeria, and Uzbekistan are significantly lower.

The results show that for each of the three considered hydrogen production sites, there is one country, which contributes most strongly to the imported social risk, whereas the imported social risk from the other involved countries is significantly lower. In the case of the Austrian production chain, this country is Uzbekistan. In the case of Germany, it is China, and in the case of Spain, this is Algeria. Furthermore, Table 10.2 also shows that the share of social risk originating from the domestic hydrogen production process in the three countries accounts for only 5.8% (Austria), 7.7% (Spain), and 8.5% (Germany), respectively. The imported social risks—the social footprint of the hydrogen production—account for more than 90% of the total social risk.

For each country, there is a further distribution of social risk. Not all stakeholders are affected equally in the respective countries from which intermediate products are imported. For the analysis, it is assumed that there are no significant differences in the main stakeholder structures along the hydrogen production chain between the countries. The first estimated distribution is also shown in Table 10.2.

In the case of the Austrian hydrogen production chain, especially, the workers in Uzbekistan carry the most considerable burden accounting for more than 11% of the total social risks, followed by the workers in China, Austria, Russia, Algeria, and Germany. In Germany, the social risks of the workers, local communities, and societies are significantly smaller. This distribution also applies to the production chains in Spain. The social risks related to the German hydrogen production chain are significantly lower for the stakeholders in the upstream production chain, whereas the social risks of the workers in Algeria in the Spanish hydrogen production chain are the second-largest (5.11) across the three analyzed production chains.

10.6 Conclusion

Using the developed SAM, we were able to show that it is possible to link the indicators of the SHDB directly to the stakeholders—the elements of the Hydrogen Economy System—of all societies around the globe connected to hydrogen production in the three countries through their specific trade relations. This multidimensional approach enables the interpretation of the top–down concepts of the Sen-Stiglitz Commission using a disaggregated social assessment model to define the social boundaries of hydrogen production.

The analysis reveals the domestic and foreign stakeholders that are affected, with regard to human well-being, by the process chains of hydrogen production in three countries, namely Austria, Germany, and Spain. Specifically, it shows that the three process chains of the countries do not affect the stakeholders in the same way. By using the developed model, various social risks associated with hydrogen production, were identified. These social risks can be interpreted as adverse external effects of hydrogen production at the three European production sites.

This social risk model thus allows the generation of detailed data about the social effects of the current—as well as changing—energy production and consumption patterns, as exemplarily shown for the case of moving towards a hydrogen economy in the given case. Coming back to the expectations of Nicolas Sarkozy, monitoring the production and consumption patterns based on the social risks model will thus show whether changes in how human well-being is measured will also change our aspirations and values.

References

Abe, J.O., Popoola, A.P.I., Ajenifuja, E., Popoola, O.M., 2019. Hydrogen energy, economy and storage: review and recommendation. International Journal of Hydrogen Energy 44 (29), 15072–15086.

Benoît Norris, C., 2014. Data for social LCA. The International Journal of Life Cycle Assessment 19, 261–265.

Benoît, C., Parent, J., Kuenzi, I., Revéret, J.P., 2007. Developing a methodology for social life cycle assessment: the North American tomato's CSR case. In 3rd International Conference on Life Cycle Management (August 02–29). Zürich, Switzerland: CIRAIG.

Benoît-Norris, C., Cavan, D.A., Norris, G., 2012. Identifying social impacts in product supply chains: overview and application of the social hotspot database. Sustainability 4 (9), 1946–1965.

Benoît-Norris, C., Norris, G.A., Aulisio Cavan, D., 2013. Social hotspots database: supporting documentation. New Earth.

Bleischwitz, R., Bader, N., Trümper, S.C., 2010. The socio-economic transition towards a hydrogen economy. Energy Policy 38 (10), 5297–5300.

BMBF (Federal Ministry of Education and Research), 2014. Research Agenda Green Economy (in German: Forschungsagenda Green Economy). BMBF, Bonn.

BMWI (Bundesministerium für Wirtschaft und Energie), 2014. Die Energie der Zukunft. Ein gutes Stück Arbeit. In: BMWI (Ed.), Erster Fortschrittsbericht zur Energiewende. BMWi, Berlin.

BMWI (Bundesministerium für Wirtschaft und Energie), 2016. The Future Energy System. Bundesministerium für Wirtschaft und Energie (BMWi), Berlin.

Campbell, W. 2014. Rare Earth Minerals: The new non-renewables. 28 February 2011. Nature blog: Nature. <http://www.nature.com/scitable/blog/green-screen/rare_earth_the_new_nonrenewables> (accessed 16.11.20.).

Cano, Z.P., Banham, D., Ye, S., Hintennach, A., Lu, J., Fowler, M., et al., 2018. Batteries and fuel cells for emerging electric vehicle markets. Nature Energy 3 (4), 279–289.

Ciroth, A., Franze, J., 2011. LCA of an ecolabeled notebook: consideration of social and environmental impacts along the entire life cycle.

Dagdougui, H., Sacile, R., Bersani, C., Ouammi, A., 2018. An overview of hydrogen economy. In: Hanane Dagdougui, R.S., Chiara, B., Ahmed, O. (Eds.), Hydrogen Infrastructure for Energy Applications Production, Storage, Distribution and Safety. Academic Press, London.

Dou, Y., Sun, L., Ren, J., Dong, L., 2017. Opportunities and future challenges in hydrogen economy for sustainable development. Hydrogen Economy. Academic Press, pp. 277–305.

Drennen, T.E., 2013. Global hydrogen resource analysis (No. SAND2013–2496P). Albuquerque, NM: Sandia National Lab.(SNL-NM).

German Federal Government, 2010. Energiekonzept für eine umweltschonende, zuverlässige und bezahlbare Energieversorgung. Berlin: German Federal Government.

German Federal Government, 2017. German Sustainability Strategy. Berlin: The German Federal Government.

German Federal Ministry of Economics and Technology (BMWI), 2012. Germany's New Energy Policy. Berlin: BMWI.

Gregory, D.P., 1973. The hydrogen economy. Scientific American 228.

Global Trade Analysis Project (GTAP), 2008. GTAP 7 Database. GTAP. Purdue University.

Hake, J.F., Fischer, W., Venghaus, S., Weckenbrock, C., 2015. The German energiewende–history and status quo. Energy 92, 532–546.

Hake, J.F., Koj, J.C., Kuckshinrichs, W., Schlör, H., Schreiber, A., Wulf, C., et al., 2017. Towards a life cycle sustainability assessment of alkaline water electrolysis. Energy Procedia 105, 3403–3410.

Hoffmann, P., 2012. Tomorrow's Energy: Hydrogen, Fuel Cells, and the Prospects for a Cleaner Planet. MIT press.

Koj, J.C., Schreiber, A., Zapp, P., Marcuello, P., 2015. Life cycle assessment of improved high pressure alkaline electrolysis. Energy Procedia 75, 2871–2877.

Koj, J.C., Wulf, C., Schreiber, A., Zapp, P., 2017. Site-dependent environmental impacts of industrial hydrogen production by alkaline water electrolysis. Energies 10 (7), 860.

Krajačić, G., Martins, R., Busuttil, A., Duić, N., da Graça Carvalho, M., 2008. Hydrogen as an energy vector in the islands' energy supply. International Journal of Hydrogen Energy 33 (4), 1091–1103.

Kuckshinrichs, W., Koj, J.C., 2018. Levelized cost of energy from private and social perspectives: the case of improved alkaline water electrolysis. Journal of Cleaner Production 203, 619–632.

Marbán, G., Valdés-Solís, T., 2007. Towards the hydrogen economy? International Journal of Hydrogen Energy 32 (12), 1625–1637.

Meadows, D.H., 2008. Thinking in Systems: A Primer. Chelsea green publishing.

Moliner, R., Lázaro, M.J., Suelves, I., 2016. Analysis of the strategies for bridging the gap towards the hydrogen economy. International Journal of Hydrogen Energy 41 (43), 19500–19508.

Moreno-Benito, M., Agnolucci, P., Papageorgiou, L.G., 2017. Towards a sustainable hydrogen economy: optimisation-based framework for hydrogen infrastructure development. Computers and Chemical Engineering 102, 110–127.

Muradov, N.Z., Veziroğlu, T.N., 2005. From hydrocarbon to hydrogen–carbon to hydrogen economy. International Journal of Hydrogen Energy 30 (3), 225–237.

Nationen, V., 1994. United Nations Development Programme (UNDP) (1994): Human Development Report. July 17, 1994.

Nations, U., 2015. Transforming Our World: The 2030 Agenda for Sustainable Development. United Nations, Department of Economic and Social Affairs, New York.

Navlani-García, M., Mori, K., Kuwahara, Y., Yamashita, H., 2018. Recent strategies targeting efficient hydrogen production from chemical hydrogen storage materials over carbon-supported catalysts. NPG Asia Materials 10 (4), 277–292.

New Earth/ Social Hotspots Database Project, 2014. Social hotspot database web portal. Introductory User Tutorial. York Beach, Maine, 03910, United States of America: New Earth. <http://socialhotspot.org/> (accessed 16.11.20.).

Newbery, D.M., 2016. Towards a green energy economy? The EU Energy Union's transition to a low-carbon zero subsidy electricity system—lessons from the UK's electricity market reform. Applied Energy 179, 1321–1330.

Norris, C.B., Norris, G.A., Aulisio, D., 2014. Efficient assessment of social hotspots in the supply chains of 100 product categories using the social hotspots database. Sustainability 6 (10), 6973–6984.

O'Garra, T., Mourato, S., Garrity, L., Schmidt, P., Beerenwinkel, A., Altmann, M., et al., 2007. Is the public willing to pay for hydrogen buses? A comparative study of preferences in four cities. Energy Policy 35 (7), 3630–3642.

Orecchini, F., 2006. The era of energy vectors. International Journal of Hydrogen Energy 31 (14), 1951–1954.

Penner, S.S., 2006. Steps toward the hydrogen economy. Energy 31 (1), 33–43.

Ren, J., Gao, S., Liang, H., Tan, S., Dong, L., 2017. The role of hydrogen energy: strengths, weaknesses, opportunities, and threats. Hydrogen Economy. Academic Press, pp. 1–33.

Rifkin, J., 2003. The Hydrogen Economy. TarcherPerigee.

Stiglitz, J.E., Sen, A., Fitoussi, J.P., 2009. Report by the commission on the measurement of economic performance and social progress.

Sonnemann, G., Valdivia, S., 2014. The UNEP/SETAC life cycle initiative. Background and Future Prospects in Life Cycle Assessment. Springer, Dordrecht, pp. 107–144.

Stiglitz, J.E., Sen, A., Fitoussi, J.P., 2010. Mismeasuring Our Lives: Why GDP Doesn't Add Up. The New Press, New York.

Unep, T., 2011. A green economy: pathways to sustainable development and poverty eradication. Nairobi Kenya Unep 12–15.

UNEP/SETA (Society of Environmental Toxicology and Chemistry C), 2009. Guidelines for Social Life Cycle Assessment of Products. UNEP, Nairobi.

United Nations, 2012. The Future We Want: Outcome Document Adopted at Rio + 20. Rio de Janeiro, UN.

van Haaster, B., Ciroth, A., Fontes, J., Wood, R., Ramirez, A., 2017. Development of a methodological framework for social life-cycle assessment of novel technologies. The International Journal of Life Cycle Assessment 22 (3), 423–440.

Venghaus, S., 2011. The management of complex system innovations. A theoretic approach to network formation and critical success factor identification using the case of fuel cell vehicles. Ph.D. Thesis. Germany: Hannover University.

Zhang, P., Guo, Y.J., Chen, J., Zhao, Y.R., Chang, J., Junge, H., et al., 2018. Streamlined hydrogen production from biomass. Nature Catalysis 1 (5), 332–338.

CHAPTER 11

Environmental impact assessments of compressed air energy storage systems: a review

Md Mustafizur Rahman, Abayomi Olufemi Oni, Eskinder Gemechu and Amit Kumar
Faculty of Engineering, Department of Mechanical Engineering, University of Alberta, Edmonton, AB, Canada

Contents

11.1 Introduction

Electricity and heat production is the most greenhouse gas (GHG)-intensive sector responsible for 31% of global emissions (Centre for Climate & Energy Solutions, 2019). Electricity demand, moreover, is expected to increase by 57% by 2050 (Bloomberg, 2018) and with it the GHG emissions' contribution. Thus, the energy sector needs deep decarbonization to meet global

Environmental Assessment of Renewable Energy Conversion Technologies
DOI: https://doi.org/10.1016/B978-0-12-817111-0.00003-6

reduction targets like the Paris Agreement, which aims to keep the global temperature rise to well below 2°C (Rogelj et al., 2016). This deep decarbonization can be done by transitioning to renewable energy. Renewable energy grew 39% between 2014 and 2018 (International Renewable Energy Agency, 2019). Among the renewable sources, solar and wind experienced huge growths of 61% and 275%, respectively, during the last five years (International Renewable Energy Agency, 2019). While these renewable sources have many environmental advantages, they also come with some challenges because of their intermittency (Hadjipaschalis et al., 2009). Increasing their shares in the grid creates stability problems, as these sources disrupt the conventional methods used in daily planning and operations (Schäfer et al., 2018). Energy storage systems (ESSs) can play an important role both in mitigating the grid stability challenges and facilitating the integration of renewables to the grid, thus ensuring deep decarbonization of the electricity sector (Beaudin et al., 2010; Arbabzadeh et al., 2019).

An ESS captures electricity from renewable and nonrenewable sources, stores it in a different form, and converts it back to electricity when required (Chen et al., 2009). Currently, the global installed ESS capacity is 181 GW, of which about 96% is pumped hydro storage (PHS) (DOE Global Energy Storage Database, 2018). Although PHS has a long life and low cost (Blakers et al., 2021), it is site-constrained (Gür, 2018), and thus its development has been limited. There is increasing interest in developing compressed air energy storage (CAES) systems (Denholm & Kulcinski, 2004). Although CAES systems are not widespread like PHSs, the technology has already proven to be mature and economically feasible (Denholm & Kulcinski, 2004). CAESs are considered prime candidates for utility-scale energy storage applications, though they have some disadvantages, such as their natural gas use and comparatively lower efficiency than PHSs (Ramadan et al., 2016). The total installed capacity of CAESs was 407 MW in 2017 (DOE Global Energy Storage Database, 2018). CAES configurations include conventional and adiabatic. In conventional compressed air energy storage (C-CAES), natural gas is combusted to produce the power to rotate the turbine. Instead of natural gas, adiabatic compressed air energy storage (A-CAES) uses thermal storage to extract heat from the compressed air, which is later used to heat the air entering the turbine. Another form of CAES is liquid air energy storage (LAES), in which the air is cooled and liquefied.

The technical characteristics and economic performance of CAES systems are well addressed in the literature. A few published articles provide information on the current development of CAESs, their technical and economic performances, and their applications and challenges. Luo et al. (2014) reviewed technical and economic characteristics, including energy and power density, specific energy, lifetime, efficiency, capital and operation costs, and more. Yu et al. (2019) reviewed the concept, classification, and technical characteristics of CAESs, and Budt et al. (2016) and Wang et al. (2017b) discussed opportunities for and obstacles to their future development.

In addition to economic and technical aspects, it is important to understand the environmental performance of CAES systems to make informed decisions for their wide deployment. To understand the GHG emission footprints, Denholm and Kulcinski (2004) evaluated the net energy ratio (NER) and life cycle GHG emissions of C-CAES. In addition to GHG emissions, Oliveira et al. (2015) estimated environmental impacts, such as human toxicity (HT), particulate matter formation (PMF), and fossil resource depletion. Neither of these studies assessed the environmental footprints of A-CAES. Bouman et al. (2016) investigated C-CAES and A-CAES systems' environmental performance for an offshore wind farm. Abdon et al. (2017) calculated the life cycle GHG emissions of C-CAES for wind and a Swiss grid mix. The studies by Bouman et al. (2016) and Abdon et al. (2017) provide useful information on material and energy inventories; however, they did not include uncertainty analyses for the data sources, assumptions, or modeling. Kapila et al. (2019) evaluated the NER and life cycle GHG emissions of C-CAES and A-CAES systems and conducted sensitivity and uncertainty analyses to capture the variability in input parameters and assumptions. The authors considered the electricity mix from several Canadian provinces for CAES charging. Electricity from renewables was not considered for charging.

All of these studies used different assumptions, material and energy inventories, and environmental impact categories. To the best of the authors' knowledge, no review assesses how life cycle assessment (LCA) has been implemented to evaluate the environmental impacts of CAES technologies. This chapter, therefore, aims at addressing this knowledge gap by compiling and reviewing all CAES-related LCA papers. Such a review provides important insight on each aspect of LCA, such as goal and scope definition, inventory analysis, environmental impact assessment, and interpretation of results.

The specific objectives are to:

- describe the concept of LCA as an environmental assessment tool;
- provide an overview of the operating principles of different CAES systems;

- perform a literature review on LCAs of CAES systems; and
- discuss how LCA is used to estimate the environmental impacts of CAES systems.

11.2 Life cycle assessment

LCA is a useful means of assessing the environmental impacts associated with a product or a system throughout its life, from raw material extraction to final disposal, often referred to as cradle-to-grave analysis. LCA originated from life cycle thinking, a way of thinking beyond the traditional focus of looking at a product system from a particular manufacturing site to include the entire life cycle and associated environmental, economic, and social impacts (UNEP-SETAC Life Cycle Initiative, 2018). LCA is an internationally standardized and harmonized tool with a set of principles and a framework (International Organization for Standardization, 2006) and requirements and guidelines (International Organization for Standardization, 2000) provided by the International Organization for Standardization (ISO). As per the ISO, the LCA framework has four phases: goal and scope definition, inventory analysis, impact assessment, and interpretation.

The first step of the LCA is defining the goal and scope. The goal is defined by stating the purposes of the LCA and identifying the intended audience. The scope comprises the functional unit, system boundary, allocation, limitations, and assumptions. The functional unit provides a reference point to which all the material and energy inputs can be related. It needs to be clearly defined in order to compare the environmental footprint of different products. The system boundary comprises the unit operations or phases considered in the study. The unit operations are often illustrated in a flow diagram that shows the product and its relationship to the unit operations.

Inventory analysis assesses the exchange between a product system and the environment. It involves collecting data and estimating the inputs (material and energy) and outputs (emissions and wastes) of each unit operation within the system boundary. A key aspect in an inventory analysis when dealing with a multifunctional process, a system with recycling or disposing of a product, is allocation. A multifunctional process is a process that yields more than one product, the desired output, and a coproduct. Most energy systems are multifunctional in nature; for instance, a cogeneration system produces electricity and heat simultaneously. The environmental burdens should be fairly attributed to both electricity and heat using the most rational allocation method. The ISO LCA standard suggests avoiding allocation whenever possible by either splitting a unit process into subprocesses and accounting for the material and energy flows for each subprocess or by expanding the system to include additional functions similar to the coproduct (International Organization for Standardization, 2000). When partitioning or system expansion is not practicable, the allocation can be done using the physical relationship of the product and its coproduct, such as mass or energy, or other relationships such as monetary values (International Organization for Standardization, 2000).

Life cycle impact assessment (LCIA) translates the emissions, natural resource use, or land use quantified in the life cycle inventory (LCI) stage to impacts on the environment. An environmental impact is any anthropogenic change to the environment, positive or negative. Environmental impacts can broadly be classified as impacts on the natural environment and human health and resource depletion (International Organization for Standardization, 2000). Resource depletion includes the depletion of mineral resources and fossil fuels (Van den Bossche et al., 2010). While climate change, ozone depletion, carcinogenic effects, and respiratory effects are included in the human health category, ecological consequences include ecotoxicity, eutrophication, acidification, and land-use change (Van den Bossche et al., 2010). LCIA lays out an environmental mechanism to scientifically characterize and measure the magnitude of the environmental impacts of each emission or resource use based on a cause-effect chain (Hauschild & Huijbregts, 2015). LCIA, according to ISO 14040/14044, has three mandatory and three optional steps. The mandatory steps are the selection and identification of impact categories, classification, and characterization. Classification is the assigning of LCI results to impact categories. For instance, carbon dioxide (CO_2), methane (CH_4), and nitrous oxide (N_2O) are assigned to "global warming." Once the LCI results are assigned to impact categories, characterization factors can be defined. Characterization factors are used to define the contribution of inventory results to the impact category. For instance, the characterization factors of CO_2 and CH_4 are 1 and 28, according to the Intergovernmental Panel on Climate Change (IPCC) Fifth Assessment Report (Myhre et al., 2013). This means that CH_4 has a global warming potential 28 times higher than CO_2 in a 100-year time horizon. The optional steps of LCIA are normalization, grouping, and weighting.

The final phase of LCA is interpretation. This involves identifying significant issues, analyzing results, drawing conclusions, and making recommendations. It also includes providing a complete and consistent presentation of results.

11.3 State-of-the-art compressed air energy storage technologies

A CAES system is a mechanical ESS that stores electricity by compressing air. The compressed air is stored in a sealed location such as a natural salt cavern, an underground rock cavern, or a high-pressure tank (Salameh, 2014; Kapila et al., 2019). Depleted oil and gas fields are considered low-cost options for storing compressed air (International Renewable Energy Agency, 2017). When there is excess electricity during a period of low demand, electricity is used to compress air to high pressure. During periods of high demand, electricity is produced from compressed air using a turbine and a generator (Breeze, 2018). A CAES system is classified as conventional or adiabatic (Kapila et al., 2019). The main components of a CAES system are the motor/generator, compressor, and recuperator (Salameh, 2014). Auxiliary equipment and an equipment control center are also part of a CAES system (Salameh, 2014).

11.3.1 Conventional compressed air energy storage

During off-peak hours, electricity is supplied to a compressor that converts the electrical energy into the potential energy of pressurized air. Usually, the air enters the compressor in atmospheric conditions. The number of compressors needed depends on the pressure at which the air is stored. This pressure is usually in the range of 45–70 bar in an underground cavern (Pollak, 1994). During on-peak hours, the air is supplied from the storage medium to a combustor where air and natural gas are combusted. Then the combustion products are released through the turbine to produce mechanical work, which is then converted to electricity using an electrical generator. Single or multiple combustors and turbines could be used depending on the power requirement. To increase the system's overall efficiency, the recuperator is placed before the combustor to preheat the air using the high-temperature exhaust stream from the turbine. Fig. 11.1 shows a schematic of a C-CAES system. The world's first C-CAES plant was built in Germany in 1978 with a capacity of 290 MW (Wang et al., 2017a). Table 11.1 shows a list of some large-scale C-CAES plants.

11.3.2 Adiabatic compressed air energy storage

In addition to the components in C-CAES, A-CAES uses thermal energy storage. Fig. 11.2 shows a schematic of an A-CAES system. Air is compressed using off-peak electricity (as it is with the C-CAES) and, in this case, stored in

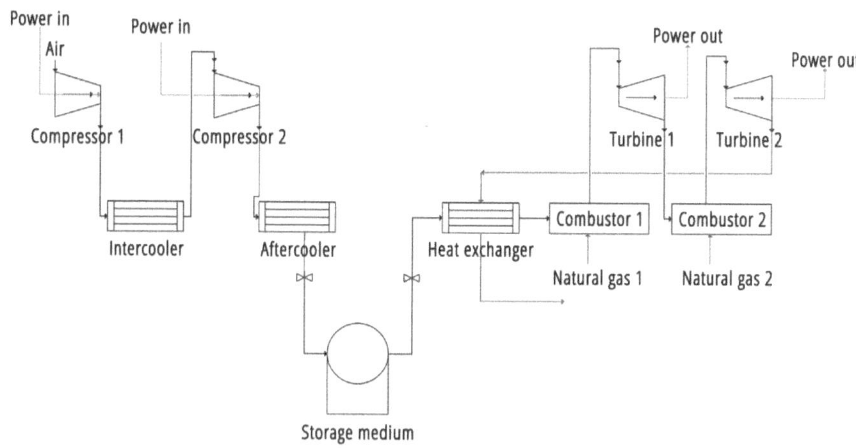

Figure 11.1 Schematic diagram of a C-CAES system.

a storage medium. The heat generated during air compression is extracted using heat exchangers and stored in a thermal fluid (molten salt solution or hot oil) for later use (Locatelli et al., 2015). After compression, the air is stored at a relatively elevated pressure of 140–160 bar in an underground cavern (Kapila et al., 2019). During on-peak periods, the heat stored in the thermal fluid is used to heat the compressed air. The hot air is then passed through the turbine to produce mechanical work, which is ultimately converted to electricity using an electrical generator. The A-CAES system does not need any additional fuel to generate electricity (Breeze, 2018). According to the DOE database, the 660 kW A-CAES plant, located in Toronto, Canada, is the only operational plant in the world (DOE Global Energy Storage Database, 2018).

11.3.3 Liquid air energy storage

LAES can be considered a form of CAES in which air is stored in a liquefied instead of a gaseous state. The energy density of the stored air is, therefore, ten times greater. A LAES system will be ten times smaller than a C-CAES system for a similar capacity plant (Breeze, 2018). Fig. 11.3 shows a simple schematic of a LAES system. The system's main components are the compressor, the air liquefaction unit, the Joule-Thomson valve, and the turbine (Highview Power, 2019). Low-cost electricity during low-demand periods is used to compress air, which is then cooled and liquefied at -196°C. The liquid air is stored in well-insulated low-pressure vessels (Highview Power, 2019) and later released and heated to regenerate gaseous air when electricity demand is high. This high-pressure air is then used directly to drive the turbine without any need for fossil

Table 11.1 List of large-scale C-CAES systems around the world.

Power rating (MW)	Location	Geological formation	Cavern volume (m^3)	Status	Source
290	Huntorf, Germany	Salt Rock	310,000	Operational	Li et al. (2018), Wang et al. (2017a)
110	McIntosh, USA	Salt Rock	580,000	Operational	Li et al. (2018), Wang et al. (2017a)
2700	Norton, USA	Hard Rock	9,600,000	Under construction	Li et al. (2018)
270	Iowa Energy Park, USA	Porous formation	-	Under construction	Li et al. (2018)

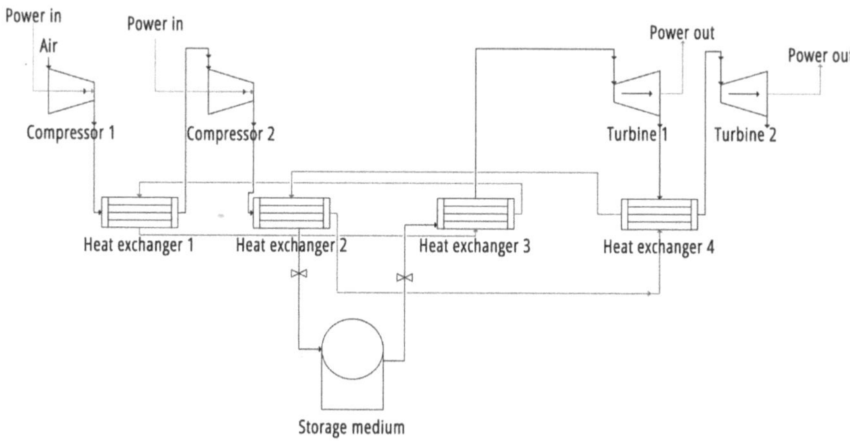

Figure 11.2 Schematic diagram of an A-CAES system.

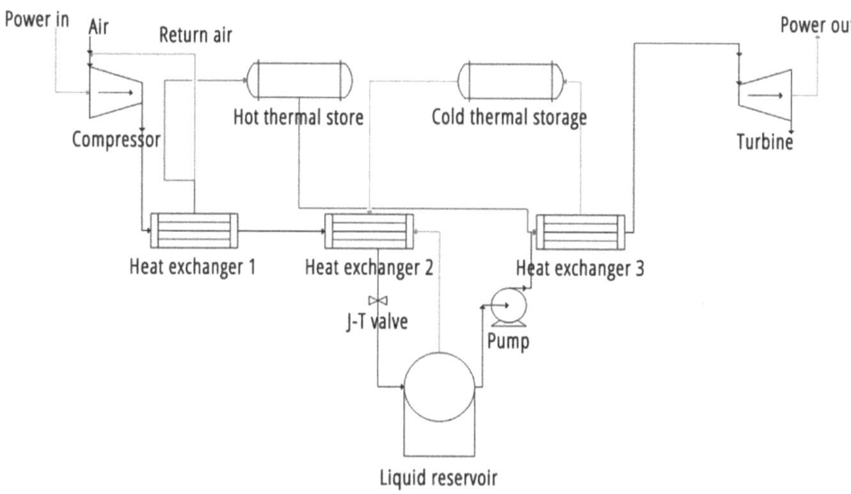

Figure 11.3 Schematic diagram of a LAES system.

fuel. To increase the overall efficiency of the system, heat from the compression of air or waste heat from industrial processes can be stored and used for heating the air after the liquefaction process (Morgan et al., 2015; Breeze, 2018). Highview Power's 5 MW/15 MWh LAES plant in the UK is the first operational demonstration of a LAES system at a grid scale. A LAES plant's overall efficiency and lifespan are 60%–75% and 30–40 years, respectively (Highview Power, 2019).

11.4 Life cycle assessment of compressed air energy storage systems

11.4.1 Overview of life cycle assessment studies on compressed air energy storage systems

Denholm and Kulcinski (2004) estimated the life cycle energy requirement and resulting GHG emissions for a 2700 MW C-CAES system for a lifetime of 40 years. The authors used process chain analysis (PCA) to quantify materials used in construction and economic input output (EIO) for manufacturing, operation, and maintenance. The life cycle energy requirement and GHG emissions were estimated as 5.27 $GJ_{thermal}$/MWh and 292 kg-CO_2equiv./MWh, respectively, without the primary electricity production. The use of natural gas in the turbine is the biggest source of GHG emissions in a C-CAES system. The construction of the cavern, site and buildings, gas infrastructure, and decommissioning together contributed less than 2% of the total GHG emissions.

Oliveira et al. (2015) evaluated the environmental performances of seven different energy storage technologies, including CAES for grid applications. The system boundary includes raw material extraction, component processing and manufacturing, distribution, product use, and end-of-life. The ReCiPe 2008 LCIA method was used to model all the technologies' midpoint category indicators (Goedkoop et al., 2009), which were later translated to a single score. SimaPro 7.3.3 was the LCA platform used. The life cycle GHG emissions were determined by the use phase, which ultimately depends on the electricity mix. The GHG emissions of CAES were 27.10 when wind energy was used in the use phase and 746.67 kg-CO_2equiv./MWh if Europe's Union for the Coordination of the Transmission of Electricity (UCTE)* 2004 electricity mix was considered. The UCTE electricity mix mainly uses fossil fuels, 30% coal and 21% natural gas. Oliveira et al. (2015) found that the life cycle GHG emissions of CAES are slightly higher than of pumped hydro but much lower than of electrochemical storage technologies.

Bouman et al. (2016) developed a process-based model to compare the cradle-to-gate environmental impacts of C-CAES and A-CAES systems. The foreground system was modeled using MATLAB combined with an Excel interface, while the data for the background system were based on the Ecoinvent v2.2. The ReCiPe v1.08's hierarchist LCIA

* UCTE coordinated the interchange of electricity in 24 countries of continental Europe till 2009.

method (Goedkoop et al., 2009) was used. The life cycle GHG emissions are 380 and 19 kg-CO_2eq/MWh for C-CAES and A-CAES, respectively. The GHG emissions of C-CAES are high due to the use of natural gas. For the C-CAES system, natural gas combustion and electricity generation from wind are the main contributors (95% of the total) for most of the impact categories. Plant construction contributes 1%–4% for most of the impact categories, except for land occupation, which makes up about 22% of construction. On the other hand, for the A-CAES system, the main contributors are thermal energy storage and electricity generation from wind, followed by construction. Due to the high material requirements, thermal energy storage makes up a large chunk of the environmental impacts. GHG emissions, photochemical oxidant formation, terrestrial acidification, and PMF are significantly lower in the A-CAES system than the C-CAES system, as no fossil fuels are combusted. The A-CAES plant requires more wind power per unit electricity delivery than the C-CAES plant, resulting in higher infrastructural requirements. As a result, impacts that are not heavily dependent on fossil fuel combustion (e.g. eutrophication, freshwater ecotoxicity, land occupation, and mineral resource depletion) increase significantly.

Kapila et al. (2019) developed process-based LCA centered on engineering first principles to assess the life cycle GHG emissions and NER of C-CAES and A-CAES systems. The NER is the ratio of total energy output to total energy input (energy required in construction, maintenance, and operation). As A-CAES does not require any natural gas during electricity production, it performs better than C-CAES for both impact categories. The NER and GHG emissions for C-CAES are 0.542 and 369 kg-CO_2equiv./MWh, respectively, for the Canada average electricity mix, and 0.702 and 232 kg-CO_2equiv./MWh, respectively, in an A-CAES system. GHG emissions are highly dominated by the operational phase, whereas construction and decommissioning have minimal impact. The GHG emissions in the operational phase for C-CAES and A-CAES are 365 and 227 kg-CO_2equiv./MWh, respectively.

Abdon et al. (2017) assessed the global warming potential of an A-CAES system. The foreground inventory data were from the literature, with some adjustments for technology performance and specification, and the background system was modeled using the ecoinvent v3.1. The authors considered the Swiss electricity mix and solar and wind power for charging in the use phase. Electricity from wind has the lowest GHG emissions (18 kg-CO_2equiv./MWh), and the electricity supply from the

Swiss grid has the highest (115 kg-CO_2equiv./MWh). The life cycle GHG emissions for a 100 MW capacity system with 4.5 hours (medium time scale) discharge ranged from 26.70 to 156.67 kg-CO_2equiv./MWh when wind and the Swiss grid mix were considered, respectively. The turbomachinery and air storage medium contribute more than 99% to the total life cycle GHG emissions for the infrastructure only, and the contribution of thermal storage is insignificant.

Sternberg and Bardow (2015) estimated the avoided GHG emissions by using 1 MWh excess electricity in C-CAES, which could be produced by conventional methods. The storage system was modeled to quantify input and output flows, which were then used to calculate the environmental impacts using the Gabi v6.3. The GHG emissions in the construction of CAES are about 5 kg-CO_2equiv./MWh. In addition, 450−470 kg-CO_2equiv./MWh are emitted during the combustion of natural gas to produce electricity. To calculate the GHGs avoided, the authors assumed electricity delivered by the ESS would replace electricity from a gas-fired power plant. The C-CAES can reduce 300−600 kg-CO_2equiv./MWh.

Bauer (2015) presented a single score using the ReCiPe endpoint assessment method for an A-CAES system. Endpoint categories, human health, ecosystems, and resources were characterized, normalized, weighted, and aggregated to get a single score of 16.8. The biggest contributor to the total environmental impact is the consumption of electricity during the use phase. The second most environmentally intensive process is compressor production. The contribution of cavern and turbine makes up less than 2% of the life cycle impact.

The studies reviewed in this chapter conducted LCA to assess various environmental impacts. These studies are different in terms of the modeling approach, data sources, and LCA platforms. There are also variations in research scope and system boundary selection. The studies did not assess the same impact categories.

The following section discusses how LCA is applied in various studies through the LCA framework. Table 11.2 gives a summary of LCAs of CAES systems.

11.4.2 Discussion on how life cycle assessment is used in compressed air energy storage studies

This section discusses how different phases of LCA are used to assess the environmental performance of CAES systems in the studies reviewed in this chapter.

Table 11.2 A list of LCAs of compressed air energy storage systems.

Author	Purpose	Functional unit	System boundary	LCIA method/Data/ Software	Impact categories	Sensitivity/ uncertainty analysis
Denholm and Kulcinski (2004)	Comparative	1 GWh	cradle-to-grave	Process chain analysis (PCA) and Economic input/ output (EIO) database	ER and GWP	N/N
Oliveira et al. (2015)	Comparative	1 kWh	cradle-to-grave	Ecoinvent and NEEDS Life Cycle Inventory Database	CC, HT, PMF, and FFD	N/N
Bouman et al. (2016)	Comparative	1 kWh	cradle-to-gate	Ecoinvent database	CC, FWET, FWE, HT, MRD, PMF, POF, TA, and AULO	N/N
Sternberg and Bardow (2015)	GHG reduction potential	1 MWh	cradle-to-gate	Ecoinvent and GaBi	GWP, FFD, HT, IR, MRD, TA, POF, PMF, and FWE	N/Y
Kapila et al. (2019)	Comparative	1 kWh	cradle-to-grave	Existing literature and ecoinvent database	NER and GWP	Y/Y

(*Continued*)

Table 11.2 (Continued)

Author	Purpose	Functional unit	System boundary	LCIA method/Data/Software	Impact categories	Sensitivity/uncertainty analysis
Abdon et al. (2017)	Comparative	1 kWh	cradle-to-grave	Existing literature and ecoinvent database	GWP	Y/N
Bauer (2015)	Comparative	1 kWh	cradle-to-gate	Data from project partners and SimaPro	CC, HT, OD, FFD, IR, AULO, PMF, FWET, etc.	N/N
Stougie et al. (2019)	Comparative	10 kWh	Cradle-to-grave	SimaPro	HT, ecosystems, and resources.	Y/N

Y, yes; *N*, no; *GWP*, global warming potential; *ER*, energy requirements; *NER*, net energy ratio; *CC*, climate change; *HT*, human toxicity; *PMF*, particulate matter formation; *FFD*, fossil fuel depletion; *FWET*, freshwater ecotoxicity; *FWE*, freshwater eutrophication; *MRD*, mineral resource depletion; *POF*, photochemical oxidant formation; *TA*, terrestrial acidification; *AULO*, agricultural and urban land occupation; *RD*, resources depletion; *OD*, ozone depletion; *IR*, ionizing radiation.

11.4.2.1 Goal and scope definition

The main purpose of a CAES system is to store excess electricity during low demand and deliver the electricity during high demand periods. Therefore, the output from the system is the amount of electricity delivered. Most LCA studies on CAES systems are, hence, aimed at evaluating and comparing the environmental advantages of CAES systems over other mechanical ESSs based on the life cycle energy and material requirement and associated emissions to deliver the same amount of electricity (Kapila et al., 2019; Denholm & Kulcinski, 2004). In addition to assessing the GHG emissions of CAES systems, a few studies identify the most influential input parameters that can affect the results and estimate a range of results by varying input parameters (Kapila et al., 2019; Abdon et al., 2017). Some studies aim to understand the GHG reduction potential when a conventional system is replaced by a CAES (Sternberg & Bardow, 2015) or compare CAES to other storage technologies such as electro-chemical batteries and fuel cell (Oliveira et al., 2015; Abdon et al., 2017). For instance, Kapila et al. (2019) developed a bottom-up LCA model to compare the environmental footprints of C-CAES and A-CAES with PHS, and Denholm and Kulcinski (2004) compared the environmental performance of C-CAES with pumped hydro and a battery storage system. Sternberg and Bardow (2015) evaluated the GHG emissions reduction when a CAES system replaces a convention gas turbine during peak load.

A functional unit is the reference against which all the inputs and outputs are normalized (Patel et al., 2016). In all the studies found on the LCA of CAESs, the unit of electricity (i.e. 1 kWh, 1 MWh, or 1 GWh) delivered from the system was used as the functional unit. Depending on the intended application and data availability, the system boundary could be defined to include the whole life cycle (i.e., cradle-to-grave) or to exclude the end-of-life phase (i.e., cradle-to-gate). However, the exclusion of life cycle stages, processes, or subprocess is acceptable only if their relative contribution to the overall environmental impact is insignificant and the reasons to omit them are justified. While most of the studies reviewed (Denholm & Kulcinski, 2004; Oliveira et al., 2015; Kapila et al., 2019; Abdon et al., 2017) considered the entire life cycle from raw material extraction to decommissioning and disposal (end-of-life), two studies did not include the end-of-life in the system boundary (Bouman et al., 2016; Sternberg & Bardow, 2015). Depending on the system boundary, the results could vary widely. For example, Denholm and Kulcinski (2004) did not consider the primary electricity generation that is used during CAES charging and estimated the GHG emissions to be only 292 kg-CO_2equiv./MWh. Nevertheless, Kapila et al. (2019) included electricity use during charging in the system boundary and estimated the GHGs to be 369 kg-CO_2equiv./MWh. In the study by Denholm and Kulcinski (2004), the location of the CAES system was unknown, and so the electricity source. However, if the source of electricity is known, along with its emission factors and system efficiency, the operational emissions can be estimated and added to the storage facility emissions to obtain the total life cycle GHG emissions. Kapila et al. (2019) included the emissions in electricity use and is specific to Canadian provinces. It is important to mention that, depending on the electricity mix, emissions from electricity use could be different, and including electricity use will result in higher overall emissions than not including it.

Another important aspect of a system boundary is the end-of-life phase. Table 11.2 shows that most of the studies included the end-of-life phase in the system boundary. Bouman et al. (2016) and Sternberg and Bardow (2015) did not consider the end-of-life phase. Although they do not give a reason, the authors may have left this phase-out because of its small contribution to the total life cycle GHG emissions or data unavailability. Depending on the system boundary definition, the life cycle GHG emissions could vary largely. Therefore, it is important to look at how the system boundary is defined when comparing different systems.

11.4.2.2 Life cycle inventory analysis

LCIA includes quantifying inputs and outputs related to the material, energy, and emissions of each unit process of a system. The reviewed studies used a number of approaches for CAES inventory analysis, including applying engineering principles and using data from project partners, existing literature, and/or LCA databases, such as the ecoinvent. Bouman et al. (2016) and Kapila et al. (2019) developed their inventories with process models using engineering principles and the ecoinvent database. In both studies, the authors used the process models to quantify the amount of electricity discharged and the storage capacity required. Bouman et al. (2016) used the ecoinvent v2.2 to calculate plant construction energy requirement, and Kapila et al. (2019) used the ecoinvent v3.3 to estimate the energy requirements for the motor/generator, compressor, and turbine.

Instead of developing a process model, some studies use published data. The LCI data in the studies by Abdon et al. (2017) and Oliveira et al. (2015) are from existing studies and the ecoinvent database. Abdon et al. (2017)

used the data from Bouman et al. (2016) for thermal storage in A-CAES and the ecoinvent v3.1 to linearly scale up the inventory data for the turbine, expander, and generator as required by the system. The inventory used by Oliveira et al. (2015) is based on research conducted at the University of Brussels. The authors used existing inventory data for the engineering components, that is, the turbine, fuel system, compressor, recuperator, etc. In addition, Oliveira et al. (2015) used the ecoinvent v2.2 for the UCTE and Belgium electricity mixes and detailed process-based inventory for material and energy for the components of the CAES system. Bauer (2015) used technical data from industry partners and applied linear scaling to generate inventory data for the turbine and compressor in SimaPro. A combination of GaBi v6.3 and ecoinvent v2.2 was used by Sternberg and Bardow (2015). They modeled the energy storage in GaBi v6.3 and used the energy requirement for plant construction from the ecoinvent v2.2 database. Denholm and Kulcinski (2004) used a hybrid approach, a PCA analysis, to assess materials and EIO to extract data for manufacturing steps and other processes. They used primary PCA for material assessments. In addition, EIO was used to get data for manufacturing steps when complete information was unavailable. Although the authors presented aggregated energy consumption in cavern development, site and building construction, and natural gas infrastructure, they did not include information on material requirements for manufacturing equipment such as the compressor, turbine, pipeline, and infrastructure (Denholm & Kulcinski, 2004).

All the studies took hybrid approaches using process models and/or existing research in combination with LCA databases such as ecoinvent or SimaPro. Developing a process model based on engineering principles requires many technical details and assumptions. Therefore, it is important to validate the model. If the information for an LCI is taken from the literature, technical specifications and performance adjustments are needed for a consistent evaluation. LCA databases like ecoinvent and SimaPro use European data. Specific energy consumption values and emission factors in these databases might not be applicable in other jurisdictions. Information taken from plant operators helps to estimate inventories more accurately but is site-specific. The data need to be adjusted to make it useful in other plants in other locations.

11.4.2.3 Life cycle impact assessment

In this phase of LCA, the inventory results are translated to impact categories, that is, global warming potential, acidification, eutrophication, fossil fuel depletion, HT, smog, land-use change, water use, etc. Global warming potential is the most common impact category in CAES LCA studies. In the reviewed studies, other impact categories were found, such as freshwater ecotoxicity, freshwater eutrophication, HT, mineral resource depletion, etc. The environmental impacts reported in different studies are presented in Table 11.3. Oliveira et al. (2015) and Bouman et al. (2016) used the ReCiPe method to calculate environmental impacts. Environmental impacts are characterized by midpoint indicators or by endpoint indicators. Midpoint indicators focus on single environmental problems, such as HT or climate change. The midpoint environmental impacts can be converted to three higher aggregation-level endpoint impacts: damage to human health, damage to ecosystems, and damage to resource availability. Oliveira et al. (2015) and Bouman et al. (2016) both estimated midpoint impact categories, and Oliveira et al. (2015) further linked the midpoint impacts to endpoint impacts and obtained a single score per kWh using midpoint-to-endpoint characterization factors. Given the global impact of climate change, most of the reviewed studies (Kapila et al., 2019; Abdon et al., 2017; Denholm & Kulcinski, 2004) emphasize global warming impact. Global warming impact is quantified by estimating GHG emissions (CH_4, N_2O, and CO_2) in CO_2 equivalents. Although it is an important impact category to help us understand the overall environmental impacts of energy consumption, a thorough assessment of mid-level impacts is necessary for making broad decisions.

11.4.2.4 Sensitivity and uncertainty analyses

Sensitivity and uncertainty analyses are not widely used in LCA studies on CAES systems. However, LCA often uses a considerable amount of data and assumptions, and these can be a source of uncertainties in the results. Although most of the reviewed studies were limited to point estimates, a few performed sensitivity and uncertainty analyses to identify the parameters that affect the results and assess the effect of a simultaneous change in multiple input parameters on the results. Such analysis is crucial, especially when the environmental impacts of different technologies are compared.

Abdon et al. (2017) conducted a sensitivity analysis using the pessimistic, expected, and optimistic values of input parameters, including lifetime, round-trip efficiency, and electricity source. Due to differences in lifetime and efficiency, the GHG emissions range from 210 to 250 kg-CO_2equiv./MWh for the short time scale. On the other hand, GHG emissions range from 75 to 225 kg-CO_2equiv./MWh because of differences in electricity type in addition

Table 11.3 A summary of environmental impacts for the C-CAES and A-CAES systems.

Technology	Author	Impact category	Value	Comment
C-CAES	Denholm and Kulcinski (2004)	Climate change	292 g-CO_2equiv./kWh	The thermal and electrical energy requirements are 5.27 MJ/kWh and 0.735 kWh/kWh, respectively. The upstream emissions for electricity generation were not considered.
	Kapila et al. (2019)	Climate change	369 g-CO_2equiv./kWh	The net energy ratio is 0.54.
	Oliveira et al. (2015)	Climate change	27.10–746.67 g-CO_2equiv./kWh	Photovoltaic, wind, the Belgium grid, and UCTE were considered for electricity generation.
		Particulate matter formation	0.08–1.0 g-PM10equiv./kWh	
		Human toxicity	24.90–475 g-1,4DBequiv./kWh	
		Fossil depletion	7.48–217 g-oilequiv./kWh	
	Bouman et al. (2016)	Climate change	380 g-CO_2equiv./kWh	The plant was modeled for 200 MW.
		Human toxicity	6.8 g-1,4DCBeqiv./kWh	
		Freshwater ecotoxicity	0.24 g-1,4DCBequiv./kWh	
		Terrestrial acidification	0.31 g-SO_2equiv./kWh	
		Particulate matter formation	0.12 g-PM10equiv./kWh	
		Mineral resource depletion	8.2 g-Fe-equiv./kWh	
	Stougie et al. (2019)	Climate change	190 g-CO_2equiv./kWh	The plant operates for 300 days/year with a lifetime of 20 years.

(*Continued*)

Table 11.3 (Continued)

Technology	Author	Impact category	Value	Comment
A-CAES	Kapila et al. (2019)	Climate change	232 g-CO_2equiv./kWh	The net energy ratio is 0.70.
	Abdon et al. (2017)	Climate change	161–247 g-CO_2equiv./kWh	The range represents the variation in lifetime and efficiency of a 100 MW system.
	Bouman et al. (2016)	Climate change	19 g-CO_2equiv./kWh	The plant was modeled for 150 MW capacity.
		Human toxicity	12 g-1,4DCBequiv./kWh	
		Freshwater ecotoxicity	0.47 g-1,4DCBequiv./kWh	
		Terrestrial acidification	0.13 g-SO_2equiv./kWh	
		Particulate matter formation	0.06 g-PM10equiv./kWh	
		Mineral resource depletion	17 g-Fe-equiv./kWh	

WebPlotDigitizer (2020) was used to generate values from graphs when required.

to lifetime and efficiency. With respect to electricity type, the authors considered wind power in Switzerland as the optimistic case and the Swiss grid mix as the pessimistic case. Although the authors conducted a sensitivity analysis, no uncertainty analysis was done to check the reliability of the results.

Kapila et al. (2019) performed a sensitivity analysis using the Sobol and Morris methods to see the impacts of flow rate, air inlet temperature, turbine inlet temperature, compressor efficiency, daily hours of operation, plant life, natural gas emission factor, and material emissions on the NER and life cycle GHG emissions. The NER and GHG emissions are most sensitive to combustion temperature and airflow rate for C-CAES and A-CAES, respectively. The authors also carried out an uncertainty analysis through a Monte Carlo Simulation to determine NER and GHG emissions variability. Due to the variability in inputs, the NERs are from 0.534 to 0.541 and 0.699 to 0.720 for C-CAES and A-CAES, respectively. On the other hand, uncertainty in GHGs ranges from 397.23 to 403.10 kg-CO_2equiv./MWh and 296.12 to 308.85 kg-CO_2equiv./MWh for C-CAES and A-CAES, respectively, when Canada's average electricity mix is used in the operation phase.

11.5 Comparison of energy storage technologies

The environmental footprint of CAES systems with other energy storage technologies was compared. The comparison is primarily based on the studies reviewed in this chapter. The studies reviewed focused mainly on mechanical ESSs, C-CAES, A-CAES, and PHS. Electrochemical batteries (lithium-ion [Li-ion], sodium-sulfur [Na-S], lead-acid [PbA], and vanadium-redox-flow-battery [VRFB]) were included with data from Hiremath et al. (2015).

11.5.1 Greenhouse gas emissions

GWP is the most widely used impact category in the reviewed studies evaluating CO_2 equivalents considering CO_2, CH_4, and N_2O emissions. Fig. 11.4 shows the GHG emissions of different ESS systems. A few studies presented only point estimates, and others provided a range of emissions. The life cycle GHG emissions are hugely influenced by the use phase (where electricity is used for charging). For instance, in A-CAES, about 98% of the total life cycle GHGs are emitted in the use phase, while the rest, 3%, are from construction and decommissioning (Kapila et al., 2019). It is important to mention that use-phase GHG emissions depend on the electricity

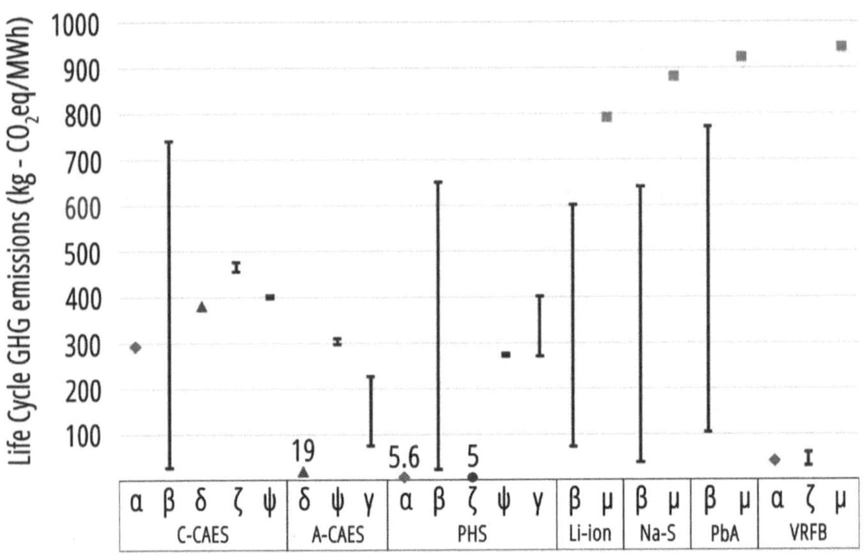

Figure 11.4 Comparison of GHG emissions for various energy storage systems.
α: Denholm and Kulcinski (2004), β: Oliveira et al. (2015), δ: Bouman et al. (2016), ζ: Sternberg and Bardow (2015), ψ: Kapila et al. (2019), γ: Abdon et al. (2017), and μ: Hiremath et al. (2015).
C-CAES, conventional compressed air energy storage; *A-CAES*, adiabatic compressed air energy storage; *PHS*, pumped hydro storage; *Li-ion*, lithium-ion; *Na—S*, sodium-sulfur; *PbA*, lead-acid; *VRFB*, vanadium-redox-flow-battery.

source. The cleaner the electricity production, the lower the life cycle GHG emissions. In Fig. 11.4, the bars for the study by Oliveira et al. (2015) represent life cycle GHG emissions ranges for different ESSs considering several sources of electricity (i.e., solar, wind, and grid mix). For example, the total GHGs for PHS ranges from 23.5 to 650 kg-CO_2equiv./MWh. While the infrastructure emissions are only about 5 kg-CO_2equiv./MWh, the GHG emissions for electricity use range from 18.5 kg CO_2equiv./MWh for the wind to 645 kg-CO_2equiv./MWh for UCTE mix. The technical parameters, such as efficiency, depth of discharge, and lifetime, also influence the total life cycle GHG emissions. For example, the life cycle GHG emissions of PHS (211 kg-CO_2equiv./MWh) are less than those of the A-CAES (232 kg-CO_2equiv./MWh) system; this is because PHS has a longer lifetime (60 years) than C-CAES (40 years) (Kapila et al., 2019).

It is clear from the above discussion that comparing only the infrastructure emissions can help understand the environmental performance of different storage technologies. Using the same system boundary for the technologies they compared, Denholm and Kulcinski (2004) estimated the GHG emissions of C-CAES, PHS, and VRFB. Electricity production emissions excluded, Denholm and Kulcinski (2004) found the highest GHG emissions in C-CAES and the lowest in PHS. While the GHG emissions for PHS and VRFB are from material and energy use in construction, the emissions in C-CAES include both construction and natural gas emissions. Of the mechanical storage systems, PHS performs better because it has a longer lifetime than other systems and does not use fossil fuels. As reported by Oliveira et al. (2015), the infrastructure emissions of electrochemical batteries are higher than those of mechanical ESSs because the batteries have a lower lifetime and use energy-intensive chemicals. The GHG emissions of C-CAES and PHS are 8 and 5 kg-CO_2equiv./MWh, respectively, while the GHG emissions of PbA, Li-ion, and Na-S are 102, 63, and 23 kg-CO_2equiv./MWh, respectively, for infrastructure only (Oliveira et al., 2015). The emissions reported by Denholm and Kulcinski (2004) and Sternberg and Bardow (2015) for VRFB are in good agreement. Although the total life cycle GHG emissions of PbA, Na–S, and Li-ion reported by Hiremath et al. (2015) are higher than those reported by Oliveira et al. (2015) due to the differences in electricity mix emissions, the GHG emissions for storage infrastructure are in good agreement.

11.5.2 Land footprint

Table 11.4 shows the land area requirement for several energy storage technologies. For electrochemical ESSs, data from Mastro (2014) were used to calculate the area required per kWh of battery capacity. The land footprint for PHS was calculated from the direct use of land and the storage capacity. For conventional and adiabatic CAESs, the values were taken from Bouman et al. (2016). The advantage of conventional electro-chemical batteries (Li-ion, Na–S, and PbA) is that they are highly modular, so they can be divided into racks and cabinets and built vertically. A VRF's land footprint is much higher than other batteries' as it requires different equipment to operate, that is, storage tanks, heat exchangers, pumps, and stacks, that are not required for Li-ion, Na-S, or PbA batteries. The main advantages of a Li-ion battery are that it is very compact and lightweight. However, the compact nature makes its weight per square meter high, which could be challenging for some building floors (Mastro, 2014). C-CAES and A-CAES systems have a large footprint because they require several components such as a compressor, turbine, heat exchanger, storage medium, etc. The A-CAES's land footprint is higher

Table 11.4 Land footprint of different energy storage technologies.

Technology	Land footprint (m^2/kWh)	Source	Comment
C-CAES	0.43	Bouman et al. (2016)	
A-CAES	0.61	Bouman et al. (2016)	
PHS	0.10	Immendoerfer et al. (2017)	Calculated from the direct use of land (m^2) and storage capacity (kWh).
	0.14	Kapila et al. (2019)	Calculated from the area of vegetation removal (m^2) and storage capacity (kWh).
Li-ion	0.02	Mastro (2014)	Calculated from the total footprint (m^2) and battery size (kWh).
Na-S	0.03	Mastro (2014)	Calculated from the total footprint (m^2) and battery size (kWh).
PbA	0.09	Mastro (2014)	Calculated from the total footprint (m^2) and battery size (kWh).
VRFB	0.24	Mastro (2014)	Calculated from the total footprint (m^2) and battery size (kWh).

than a C-CAES's because it needs storage for the thermal fluid that extracts heat during air compression and rejects heat to increase air temperature before the air enters the turbine, in addition to the other components (Kapila et al., 2017). For PHS, the land footprint ranges from 0.10 m^2 (Immendoerfer et al., 2017) to 0.14 m^2/kWh (Kapila et al., 2019).

11.6 Conclusions and recommendations

With increasing renewables shares, ESSs are needed to handle the variability and unpredictability in electrical grids. Increased environmental concern is behind the push to quantify the environmental footprints of ESSs. CAES, a leading storage technology, has already attracted considerable attention from researchers. To evaluate environmental footprints, researchers use LCA. This chapter reviewed several LCA studies on CAES systems and discussed how LCA is applied to evaluate environmental impacts. This review identified some gaps. Although there are a few LCAs of C-CAES and A-CAES, no LCA was found for the emerging technology, LAES. A LAES system uses electricity to liquefy air that can be stored in a well-insulated tank, and it can be built anywhere. An in-depth analysis of its environmental performance is required, and future LCAs should include it. Most of the studies reviewed here address GHG emissions only. Other environmental impacts, that is, acidification, land use, water use, freshwater ecotoxicity, freshwater eutrophication, HT, etc., are not well covered.

Future LCA studies should focus on using appropriate values for emission factors and specific energy consumption. While conducting an LCA, one must deal with a vast quantity of data and many assumptions, making the results uncertain. In earlier studies, little attention was given to uncertainty analysis. This area also requires further study.

Acknowledgments

The authors thank the NSERC Energy Storage Technology (NEST) Network (RYERU NSERC 468468 Kumar) and the University of Alberta (UOFAB VPRGRF NEST KUMAR) for the financial support provided to carry out this research. We would also like to thank the NSERC/Cenovus/Alberta Innovates Associate Industrial Research Chair Program in Energy and Environmental Systems Engineering and the Cenovus Energy Endowed Chair in Environmental Engineering at the University of Alberta for financial support for this study. We also thank Astrid Blodgett for editing this chapter.

References

Abdon, A., Zhang, X., Parra, D., Patel, M.K., Bauer, C., Worlitschek, J., 2017. Techno-economic and environmental assessment of stationary electricity storage technologies for different time scales. Energy 139, 1173–1187.

Arbabzadeh, M., Sioshansi, R., Johnson, J.X., Keoleian, G.A., 2019. The role of energy storage in deep decarbonization of electricity production. Nature communications 10 (1), 1–11.

Bauer, C., 2015 Screening of a lifecycle Assessment (LCA) of an Advanced Adiabatic Compressed Air Energy Storage.

Beaudin, M., Zareipour, H., Schellenberglabe, A., Rosehart, W., 2010. Energy storage for mitigating the variability of renewable electricity sources: An updated review. Energy for sustainable development 14 (4), 302–314.

Bloomberg, N.E.F., 2018. Global electricity demand to increase 57% by 2050. <https://about.bnef.com/blog/global-electricity-demand-increase-57-2050/> (accessed 03.10.2019).

Blakers, A., Stocks, M., Lu, B., Cheng, C., 2021. A review of pumped hydro energy storage. Progress in Energy 3022003.

Bouman, E.A., Øberg, M.M., Hertwich, E.G., 2016. Environmental impacts of balancing offshore wind power with compressed air energy storage (CAES). Energy 95, 91–98.

Breeze, P., 2018. Chapter 3 - Compressed air energy storage. In: Breeze, P. (Ed.), Power System Energy Storage Technologies. Academic Press, United Kingdom.

Budt, M., Wolf, D., Span, R., Yan, J., 2016. Compressed air energy storage—an option for medium to large scale electrical-energy storage. Energy procedia 88, 698–702.

Centre for Climate and Energy Solutions. 2019. *Global Emissions*. <https://www.c2es.org/content/international-emissions/#targetText = Notes,%25)%20and%20manufacturing%20(12%25> (accessed 03.10.2019).

Chen, H., Cong, T.N., Yang, W., Tan, C., Li, Y., Ding, Y., 2009. Progress in electrical energy storage system: A critical review. Progress in natural science 19 (3), 291–312.

Denholm, P., Kulcinski, G.L., 2004. Life cycle energy requirements and greenhouse gas emissions from large scale energy storage systems. Energy Conversion and Management 45 (13–14), 2153–2172.

DOE Global Energy Storage Database. 2018. <http://www.energystorageexchange.org/projects/data_visualization> (accessed 03.07.2018).

Goedkoop, M., Heijungs, R., Huijbregts, M., De Schryver, A., Struijs, J., Van Zelm, R., 2009. ReCiPe 2008, A Life Cycle Impact Assessment Method Which Comprises Harmonised Category Indicators at the Midpoint and the Endpoint Level. Ministry of Housing, Spatial Planning and Environment, pp. 1–126.

Gür, T.M., 2018. Review of electrical energy storage technologies, materials and systems: challenges and prospects for large-scale grid storage. Energy & Environmental Science 11 (10), 2696–2767.

Hadjipaschalis, I., Poullikkas, A., Efthimiou, V., 2009. Overview of current and future energy storage technologies for electric power applications. Renewable and Sustainable Energy Reviews 13 (6–7), 1513–1522.

Hauschild, M.Z., Huijbregts, M.A., 2015. Introducing life cycle impact assessment. In Life cycle impact assessment. Springer, Dordrecht, pp. 1–16.

Highview Power. 2019. The energy we store can power up to 200,000 houses for a whole day. <https://www.highviewpower.com/technology/> (accessed 04.10.2019).

Hiremath, M., Derendorf, K., Vogt, T., 2015. Comparative life cycle assessment of battery storage systems for stationary applications. Environmental Science & Technology 49 (8), 4825–4833.

Immendoerfer, A., Tietze, I., Hottenroth, H., Viere, T., 2017. Life-cycle impacts of pumped hydropower storage and battery storage. International Journal of Energy and Environmental Engineering 8 (3), 231–245.

International Organization for Standardization 2000. ISO 14042:2000, Environmental management—life cycle assessment—life cycle impact assessment. International Organization for Standardization.

International Organization for Standardization 2006. ISO 14040: 2006, Environmental management—Life cycle assessment—Principles and framework. International Organization for Standardization.

International Renewable Energy Agency. 2017. Electricty storage and renewables: costs and markets to 2030. <http://www.irena.org/publications/2017/Oct/Electricity-storage-and-renewables-costs-and-markets> (accessed 04.10.2019).

International Renewable Energy Agency. 2019. Renewable Energy Statistics 2019. <https://www.irena.org/-/media/Files/IRENA/Agency/Publication/2019/Jul/IRENA_Renewable_energy_statistics_2019.pdf> (accessed 23.09.2019).

Kapila, S., Oni, A.O., Kumar, A., 2017. The development of techno-economic models for large-scale energy storage systems. Energy 140, 656–672.

Kapila, S., Oni, A.O., Gemechu, E.D., Kumar, A., 2019. Development of net energy ratios and life cycle greenhouse gas emissions of large-scale mechanical energy storage systems. Energy 170, 592–603.

Li, L., Liang, W., Lian, H., Yang, J., Dusseault, M., 2018. Compressed air energy storage: characteristics, basic principles, and geological considerations. Advances in Geo-Energy Research 2 (2), 135–147.

Locatelli, G., Palerma, E., Mancini, M., 2015. Assessing the economics of large Energy Storage Plants with an optimisation methodology. Energy 83, 15–28.

Luo, X., Wang, J., Dooner, M., Clarke, J., Krupke, C., 2014. Overview of current development in compressed air energy storage technology. Energy Procedia 62, 603–611.

Mastro, C. 2014. Behind-the-meter battery storage: Technical and market assessment. Report Number 15-02. New York State Energy Research and Development Authority. Albany, NY.

Morgan, R., Nelmes, S., Gibson, E., Brett, G., 2015. An analysis of a large-scale liquid air energy storage system. Proceedings of the Institution of Civil Engineers-Energy 168 (2), 135–144.

Myhre, G, Shindell, D, Bréon, F, Collins, W, Fuglestvedt, J, Huang, J, et al., 2013. Anthropogenic and natural radiative forcing, climate change 2013: The Physical Science Basis. Contribution of Working Group I to the Fifth Assessment Report of the Intergovernmental Panel on Climate Change. Cambridge University Press, Cambridge, pp. 659–740.

Oliveira, L., Messagie, M., Mertens, J., Laget, H., Coosemans, T., Van Mierlo, J., 2015. Environmental performance of electricity storage systems for grid applications, a life cycle approach. Energy Conversion and Management 101, 326–335.

Patel, M., Zhang, X., Kumar, A., 2016. Techno-economic and life cycle assessment on lignocellulosic biomass thermochemical conversion technologies: A review. Renewable and Sustainable Energy Reviews 53, 1486–1499.

Pollak, R. 1994. History of first US compressed-air energy storage (CAES) Plant (110MW 26h) volume 2: Construction. Electric Power Research Institute (EPRI).

Ramadan, O., Omer, S., Jradi, M., Sabir, H., Riffat, S., 2016. Analysis of compressed air energy storage for large-scale wind energy in Suez, Egypt. International Journal of Low-Carbon Technologies 11 (4), 476–488.

Rogelj, J., Elzen, Den, Höhne, M., Fransen, N., Fekete, T., Winkler, H., et al., 2016. Paris Agreement climate proposals need a boost to keep warming well below 2 C. Nature 534 (7609), 631–639.

Salameh, Z., 2014. Chapter 4 - Energy storage. In: Salameh, Z. (Ed.), Renewable Energy System Design. Academic Press, Boston.

Schäfer, B., Beck, C., Aihara, K., Witthaut, D., Timme, M., 2018. Non-Gaussian power grid frequency fluctuations characterized by Lévy-stable laws and superstatistics. Nature. Energy 3 (2), 119–126.

Sternberg, A., Bardow, A., 2015. Power-to-What?–Environmental assessment of energy storage systems. Energy & Environmental Science 8 (2), 389–400.

Stougie, L., Del Santo, G., Innocenti, G., Goosen, E., Vermaas, D., Van der Kooi, H., et al., 2019. Multi-dimensional life cycle assessment of decentralised energy storage systems. Energy 182, 535–543.

UNEP-SETAC Life Cycle Initiative. 2018. What is life cycle thinking? <https://www.lifecycleinitiative.org/starting-life-cycle-thinking/what-is-life-cycle-thinking/> (accessed 20.12.2019).

Van den Bossche, P., Matheys, J., Van Mierlo, J., 2010. Battery Environmental Analysis. Electric and Hybrid Vehicles: Power Sources, Models, Sustainability, Infrastructure and the Market. Elsevier, Netherland, p. 347.

Wang, J., Lu, K., Ma, L., Wang, J., Dooner, M., Miao, S., et al., 2017a. Overview of compressed air energy storage and technology development. Energies 10 (7), 991.

Wang, J., Ma, L., Lu, K., Miao, S., Wang, D., Wang, J., 2017b. Current research and development trend of compressed air energy storage. Systems Science & Control Engineering 5 (1), 434–448.

WebPlotDigitizer. 2020. <https://automeris.io/WebPlotDigitizer/> (accessed 26.11.2020).

Yu, Q., Wang, Q., Tan, X., Fang, G., Meng, J., 2019. A review of compressed-air energy storage. Journal of Renewable and Sustainable Energy 11 (4), 042702.

CHAPTER 12

Environmental impact assessment of battery storage

M. A. Parvez Mahmud[1] and Nahin Tasmin[2]
[1]School of Engineering, Deakin University, Geelong, VIC, Australia
[2]Department of Mechanical Engineering, Rajshahi University of Engineering & Technology, Kazla, Rajshahi, Bangladesh

Contents

12.1 Introduction

Battery storages are essential to operate electrical and electronic devices or machines. They provide continuous backup to the main grid power supply to run the industries all over the world. Electric vehicles are dependent on batteries. The small electronic devices are also fully reliant on batteries. Thus, millions of batteries are globally used in our everyday life to make devices operational. The use of these batteries will increase in the future as people are getting used to more and more electronic devices as well as electric machines. However, these battery storages have substantial environmental impacts due to the used chemicals (DOE Global Energy Storage Database, 2017). Many research studies were conducted to check the environmental profiles of batteries considering a part of their life. There is still

Environmental Assessment of Renewable Energy Conversion Technologies
DOI: https://doi.org/10.1016/B978-0-12-817111-0.00001-2

limited research carried out on life cycle analysis of battery storages like lithium-ion (Li-ion), sodium chloride (NaCl), and nickel-metal hydride (NiMH). Therefore, this research considered these batteries and assessed their environmental effects by considering their entire life cycle.

The environmental impact evaluation through life cycle assessment (LCA) is an arduous job. It involves the effects from the production of the elements at whole lifetime that are raw material extraction to the end of life recycling (IEA, 2016). At first, a considerable literature review was conducted considering keywords LCA, environmental impact, Li-ion, NaCl, NiMH, and greenhouse gas (GHG) emissions. It helped to gather hazardous emission rates from the battery storage in the overall lifetime. The evaluation of used LCA methods, data sources, and simulation tools helped to design the research methodology of this work. It is necessary to estimate fossil-fuel-based energy consumptions to replace them with renewable sources. Moreover, the assessment of GHG emissions to air for various battery parts and life stages is essential to find out the impacting elements. Therefore, it is also necessary to replace the dangerous elements with environmental-friendly options. Besides, it is necessary to compare the lifetime environmental impacts of Li-ion, NaCl, and NiMH battery storage to discover the best option.

LCA is a well-known state-of-the-art and effective approach to evaluate the environmental effects of a unit process or system. It considers the impacts of many environmental indicators such as carcinogens, non carcinogens, respiratory inorganics, ionizing radiation, ozone layer depletion, aquatic ecotoxicity, land occupation, mineral extraction, nonrenewable energy, terrestrial ecotoxicity, respiratory organics, and global warming. It also provides the effects for the end-point categories of human health, ecosystem quality, climate change, and resources. Moreover, it reveals the overall metal emissions such as mercury, zinc, cadmium, lead, and copper. Furthermore, it gives the estimation of GHG emissions such as carbon dioxide, nitrogen oxide, sulfur oxide, and ammonia, considering the whole timespan. Finally, it assists in comparing the effects of Li-ion, NaCl, and NiMH battery storage. During the impact assessment by LCA, all the input and output material-flows and emissions at every life stage of the considered battery storages are taken into account (Stavropoulos et al., 2016; Finnveden et al., 2009; Curran, 2013). It also considers the effects from the extraction of raw materials to the manufacturing, operation, and after-life waste disposal (Židonienė & Kruopienė, 2015; Mahmud et al., 2018a,b).

The previous LCA research on LCA of Li-ion battery storage conducted by Liang et al. (2017) depicted that the maximum amount of GHG emissions took place during the productions and recycling stages (Liang et al., 2017). They estimated and compared the carbon footprints of three cases of Li-ion battery storage. Nevertheless, their study did not account for the geographical effect discrepancies. Innocenzi et al. showed the state-of-the-art LCA process and treatment techniques of NiMH battery storage (Innocenzi et al., 2017). They exemplified the NiMH collectors and assessed the end-of-life disposal routes. The research works conducted by Notter et al. (2010) and Ellingsen et al. (2014) estimated the effects of Li-ion battery storage. However, the exact direct impacts on humans, ecosystems, and resources have not been evaluated. Meng et al. (2017) assessed the nonaqueous electrolytes of NiMH battery storage considering ionic liquid. For that purpose, they conducted the charging and discharging experimentations. They also used acid mixtures to replace the electrolytes like potassium hydroxide (KOH). Nevertheless, the effects of these elements on the environment have not been assessed. Hao et al. (2017) qualified the GHG emissions at the Li-ion battery parts manufacturing. They did not consider the whole life stages of the storage. According to the authors' experience, LCA studies of battery storage considered only one life cycle stage, as the manufacturing or recycling phase. A complete cradle-to-grave-based LCA analysis of Li-ion, NaCl, and NiMH battery storage has not been conducted yet, which is essential to realize the best option. Also, it is necessary to replace the impacting elements, devices, or processes with an equivalent environment-friendly item. Therefore, this research aims to calculate the environmental impacts of Li-ion, NaCl, and NiMH battery storage and compare them considering the environmental impact assessing categories mentioned above.

The new comprehensive life cycle inventories (LCI) are developed for all three battery storages. The prior relevant LCA studies and ecoinvent database helped to construct the LCI through arranging all inputs and outputs of the life cycle (Baj, 2017). It also assisted in developing the LCA system boundary of the considered battery storages (Battery Uni) (Barry Huret, president of battery consulting company Huret Associates Inc. in Yardley, PA, USA). The SimaPro software version 8.5 was used to estimate and compare the effects in terms of humankind, resources, and ecosystem. The Impact 2002 + , EcoPoints 97, and cumulative energy demand (CED) methods were utilized for assessing the overall impacts of the battery storage. The main contributions of this research are outlined below:

- New comprehensive LCI formation for Li-ion, NaCl, and NiMH battery storage.
- Environmental profiles of these storages' assessment considering total lifetime.
- Environmental impacts of the considered storage comparison and determining the best option in terms of fewer emissions and reduced fossil-fuel-based energy consumptions.
- Metal- and gas-based effects of the battery storages to humankind, the ecosystem, and resources were evaluated.

Based on the above discussion, the rest of the article is organized as follows. Section 2 describes markets of battery storage and production overview. The research approach of this LCA study is highlighted in Section 12.3. Section 12.4 depicts the results by the Impact 2002 + , EcoPoints 97, and CED methods. The explanation of findings and their comparisons are discussed in Section 12.5. Section 12.6 depicts the constraints of this work. In the end, the concluding summary is given in Section 12.7.

12.2 Battery storage markets and production overview

The need for battery storage is enormous to store power and supply it during main grid cut-offs worldwide. Moreover, battery storage is widely used to run portable electronics continuously. Day-by-day, sensors, actuators, personal electronics, and biomedical devices are getting popular to make life more convenient. Therefore, the usage of battery storage is augmenting—many countries producing rechargeable and nonrechargeable battery storages to fulfill the ever-increasing demands. The battery storage markets and production overview are highlighted in Fig. 12.1.

It is seen that the Li-ion market was approximately US\$37 billion in 2012, which is the second maximum among all battery storages (Battery Uni) (The Freedonia Group, Inc.). The necessity to use Li-ion battery storage had augmented approximately seven times in 15 years period. However, the NiMH battery storage demand was approximately US\$9 billion in 2012. The trend analysis shows that the use of battery storage is augmenting over time. Fig. 12.2 depicts that Japan is the maximum battery-producing nation in the world (Battery Uni).

Fig. 12.3 presents an overview of the recent rechargeable battery production in Japan, collected from the Total Battery Production Statistics of Japan in the year 2017 (Baj, 2017). As can be seen, in total,

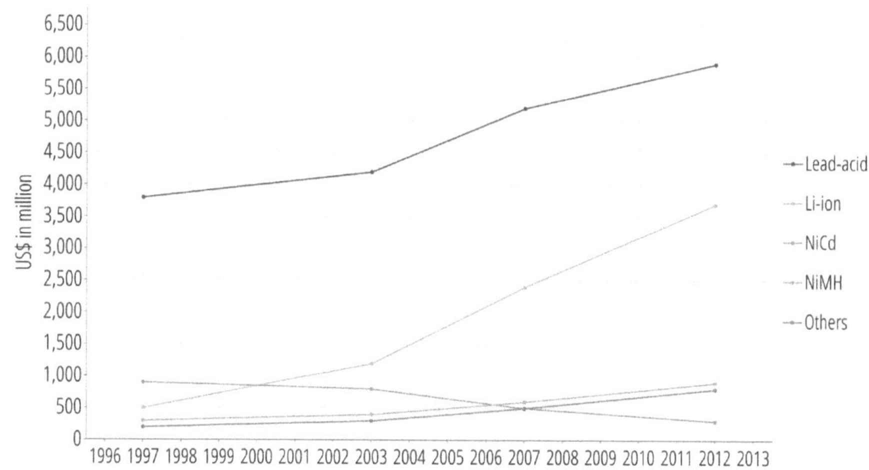

Figure 12.1 Global demands of rechargeable batteries.

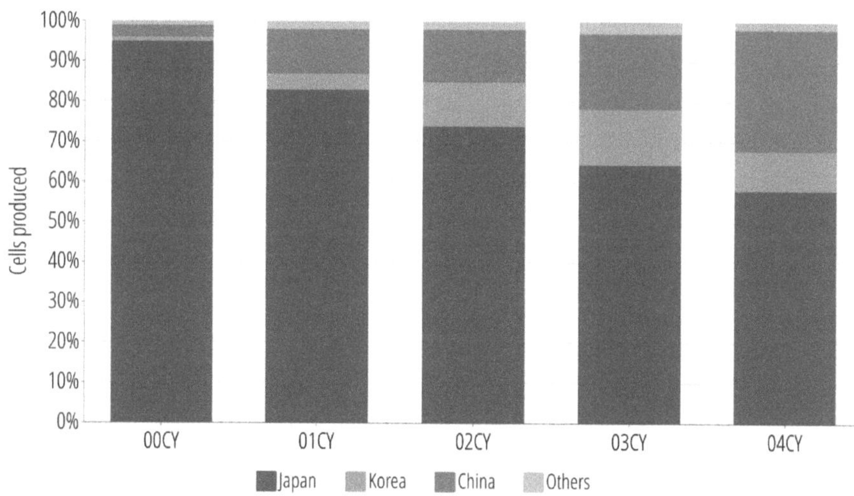

Figure 12.2 Productions of rechargeable batteries in the countries (Battery Uni).

around 4.14 billion batteries were produced with the following shares: 43% rechargeable batteries with 31% (by volume) being Li-ion type and 10% of the NiMH type. From Fig. 12.3, it can also be depicted that around 92% of the total production value (814.1 billion JPY) was

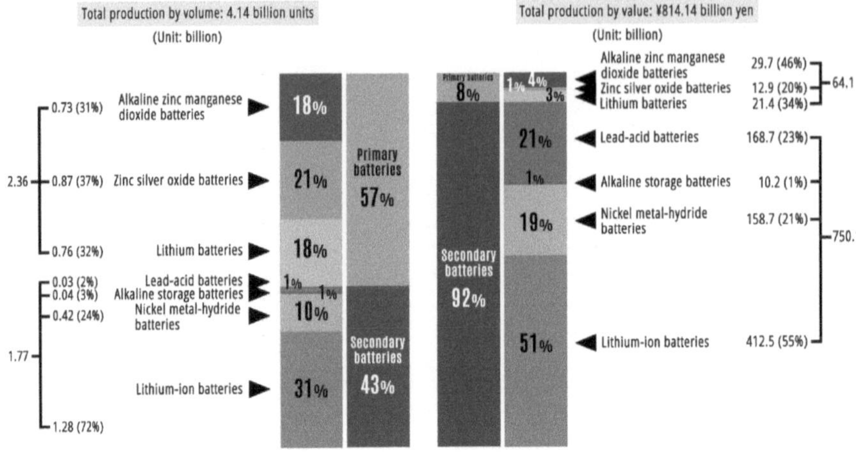

Figure 12.3 Rechargeable battery production statistics in Japan (Baj, 2017).

responsible for secondary battery production. Of the secondary battery production values, about 51% were Li-ion batteries and 19% NiMH batteries. Hence, the entire outline of contemporary battery production shows that the production rate of Li-ion and NiMH batteries is much higher compared to other rechargeable battery production in Japan.

Table 12.1 groups the selling history of Li-ion and NiMH batteries in Japan from 2013–17, and the data was arranged from the Battery Association of Japan (2018). As presented, the selling price of Li-ion batteries was around 279.364 million JPY in the year 2013, while at the same time, the selling price for NiMH batteries was about 178.748 million JPY. However, the demand for Li-ion batteries augmented per year with an opposite scene in the case of NiMH batteries. Over five years, the selling price of Li-ion batteries was about 422.366 million JPY, and that of NiMH batteries was around 166.583 million. Also, both Li-ion and NiMH batteries showed similarities in assessing the total number of sales over these time periods (2013–17). The huge selling of these batteries also points to environmental impacts to a greater degree. Hence, it is essential to identify environmentally hazardous chemicals or apparatus from Li-ion and NiMH batteries to replace them with an environment-friendly equivalent substitute without affecting the performance.

Table 12.1 Rechargeable battery sale statistics in Japan.

	Amount (million JPY)				
	2013	2014	2015	2016	2017
Li-ion	279,364	349,761	360,705	385,368	422,366
NiMH	178,748	168,157	160,294	167,785	166,583
	Quantity (1000 pcs)				
	2013	2014	2015	2016	2017
Li-ion	844,622	955,644	1,031,850	1,277,108	1,320,502
NiMH	455,132	432,936	434,397	478,210	471,105

12.3 Methodology

This paper analyzes and compares Li-ion's life cycle environmental impact and impact assessment technology based on ISO standardized methods. The LCA methodology is formed based on ISO 14040:2006 to ISO 14044:2006 (Pryshlakivsky & Searcy, 2013; Mahmud et al., 2018d). The datasets are collected from the ecoinvent database (Pascual-González et al., 2016; Mahmud et al., 2016), where these are originated from Hischier et al. (2007).

Four basic steps should be followed to perform a complete LCA: goal and scope definition, life cycle inventory, life cycle impact assessment, and results from interpretation (Farjana et al., 2018a; Parvez Mahmud et al., 2018; Mahmud & Farjana, 2012). The first step of LCA is the goal and scope definition. This study aims to analyze and compare the environmental impacts caused by Li-ion and NiMH batteries based on standardized methods. The scope of this research is the environmental emissions that are generated from the production of the unit amount of these batteries. The specifications of the considered batteries for life cycle environmental impact assessments are given in Table 12.2 (Hischier et al., 2007). Figs. 12.4 and 12.5 show the material flows in various stages of the Li-ion and NiMH battery plants, respectively.

The types of energy consumption during the raw material extraction to Li-ion battery production are low voltage (0.81%), medium voltage (15.7%), and high voltage (16%). Biomass energy from sugarcane is consumed mostly with a rate of 35.4% throughout the life cycle of the Li-ion battery. About 30.1% of medium electricity is used for aluminum production. The respective rates of used aluminum and copper are 32.8% and 35.5%.

NiMH batteries mostly use nickel at a rate of approximately 74.7%. About 60.5% of the Ni is used for the negative electrode, whereas about

Table 12.2 Specifications of Li-ion and NiMH batteries.

	Li-ion	NiMH
Model	FPCBP64	UBP0767B
Size [1 × w × h(mm)]	128.00 × 77.00 × 20.00	138.00 × 105.00 × 20.00
Weight (kg)	0.19	0.591
Capacity (mAh)	4000	4500
Number of cells	4	1

Source: Adopted from Hischier R., Classen M., Lehmann M., and Scharnhorst W., 2007, Life cycle inventories of electric and electronic equipment: production, use and disposal. Swiss Centre for Life Cycle Inventories, United Kingdom.

23.5% is used for the battery's positive electrode. The rates of other metals and earth substances are approximately 16.1% and 8.16%, respectively. Like Li-ion batteries, NiMH also uses low, medium, and high voltage electricity at different stages of life. The rates of using low, medium, and high voltages are 10.1%, 9.52%, and 9.88%. However, NiMH consumes about 9.26% of the electricity mix originating from primary and secondary sources. About 6.67% of low-voltage electricity consumption is from sugarcane-based biomass plants.

The functional unit (Lewandowska et al., 2011; Farjana et al., 2018c; Mahmud et al., 2018c) is 1 kg of battery produced of each type (Li-ion and NiMH batteries). The system boundary (Fig. 12.6) here includes the battery production phase, cradle-to-gate (Stavropoulos et al., 2016; Farjana et al., 2018b,d). The materials, forms of energies in renewables and nonrenewables, transportation are considered as material inputs here. The output is 1 kg of battery produced for both types and the heat waste emissions. The LCIof the Li-ion and NiMH battery plants are highlighted in Tables 12.3 and 12.4, respectively. This analysis is done using SimaPro software version 8.5 (Marsmann & Schiburr, 1993). The geographic region under consideration is taken as in Japan. It is assumed that the battery plant is in Japan, and the raw materials used and the electricity in use are produced in the Asian continent. As in battery production, no co-product is produced in conjunction with batteries, and there is no need to apply allocation techniques.

For assessing the impacts, several LCA methodologies are applied, which made it a hybrid LCA. The methodologies discussed here are CML, ReCiPe, EcoPoints, IPCC, raw material flow, and CED method. The raw material flow method analyses and quantifies the emissions based on raw material inputs and outputs emitted to the soil, water, and air and tracks the

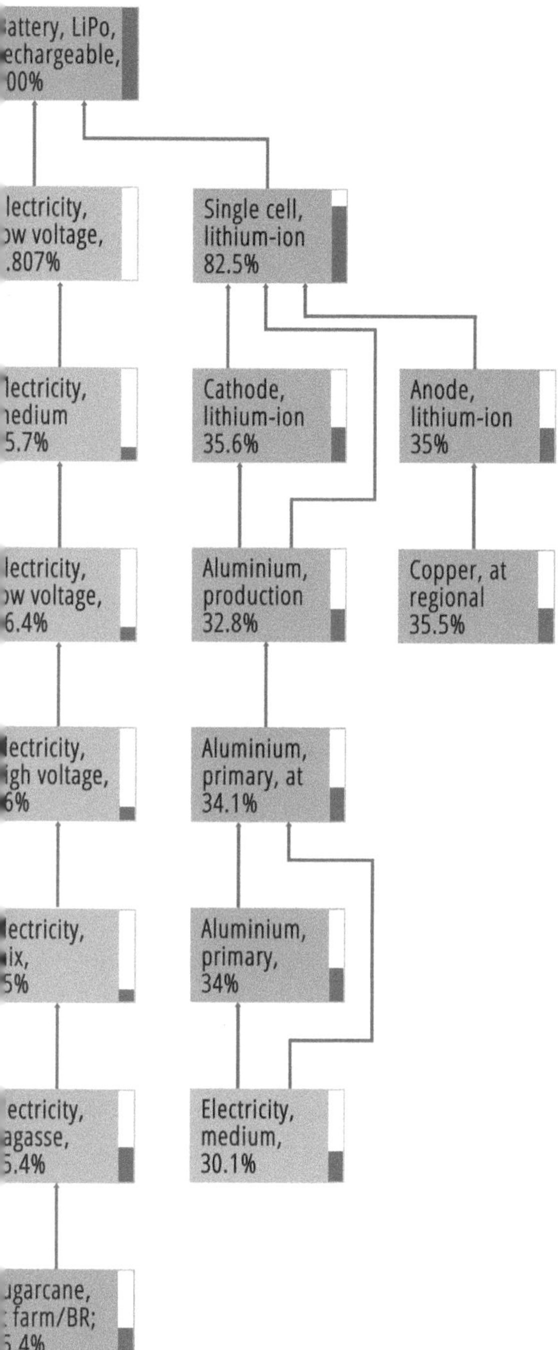

ure 12.4 Material flow sheet for 1 kg of Li-ion battery duction. *From Parvez Mahmud, M.A., Huda, N., Farjana, S. Lang, C., 2019. A strategic impact assessment of hydro-ver plants in alpine and non-alpine areas of Europe, Applied rgy, 250, 198-214. https://doi.org/10.3390/batteries5010022.*

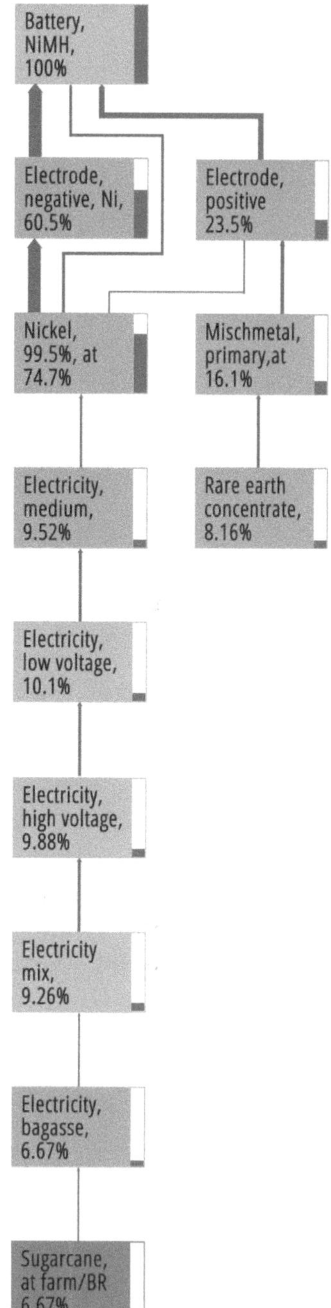

Figure 12.5 Material flow sheet for 1 kg of NiMH battery production. *From Parvez Mahmud, M.A., Huda, N., Farjana, S.H., Lang, C., 2019. A strategic impact assessment of hydropower plants in alpine and non-alpine areas of Europe, Applied Energy, 250, 198-214. https://doi.org/10.3390/batteries5010022.*

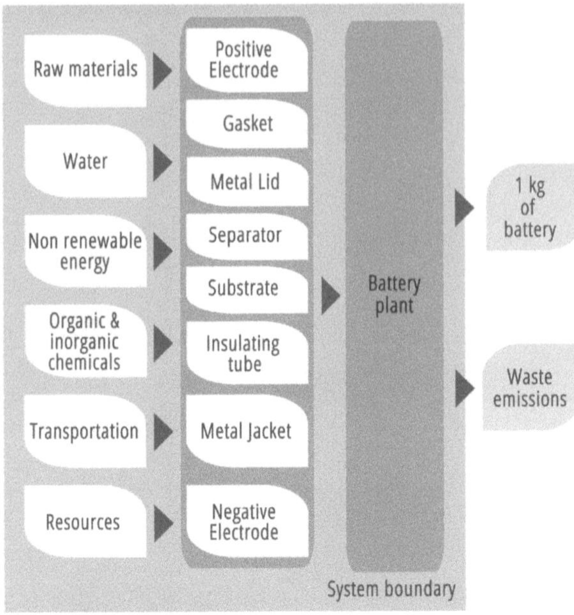

Figure 12.6 LCA system boundary. *From Parvez Mahmud, M.A., Huda, N., Farjana, S.H., Lang, C., 2019. A strategic impact assessment of hydropower plants in alpine and non-alpine areas of Europe, Applied Energy, 250, 198-214. https://doi.org/10.3390/batteries5010022.*

plant emissions. On the other hand, the CML method and ReCiPe method are specifically designed to analyze the production processes or systems of processes located in the European sub-continent. The CML method (Ligthart et al., 2010), developed by the Center for Environmental Sciences of Leiden University, aims to analyze the environmental effects based on 10 impact categories. These categories include global warming, human toxicity, abiotic depletion, ecotoxicity, terrestrial ecotoxicity, eutrophication, marine aquatic ecotoxicity, and photochemical oxidation. The ReCiPe method (Audenaert et al., 2012) also analyses the environmental impact based on different categories, either midpoint-based or endpoint-based.

The intergovernmental panel of climate change (IPCC) method (Minx et al., 2017) aims to analyze the GHG emissions based on different time frames like 20 years, 50 years, or 100 years. The EcoPoints method (Pesso, 1993) analyses the emissions based on 13 impact categories based on major elements emitted to the environment: carbon dioxide (CO_2), nitrogen oxide (NO_x), sulfur oxide (SO_x), ammonia (NH_3), nitrogen (N), copper (Cu), cadmium (Cd), Nickel (Ni), phosphorous (P), zinc (Zn), mercury (Hg), lead (Pb), and chromium (Cr).

Table 12.3 Life cycle inventory of Li-ion battery.

Products	**1.00**	**kg**
Battery, Li-ion, rechargeable, prismatic, at plant/GLO U/AusSD U		
Inputs		
Reinforcing steel	1.45E-01	kg
Sheet rolling, steel	1.45E-01	kg
Transport, transoceanic freight ship	7.81	tkm
Electricity, low voltage, production UCTE, at grid	1.08E-01	kWh
Transport, lorry >16t	1.02	tkm
Cable, three-conductor cable	2.50E-02	m
Cable, data cable in infrastructure	3.73E-01	m
Printed wiring board, surface mounted, unspec., solder mix	3.38E-03	kg
Single-cell, Li-ion battery, lithium manganese oxide/graphite	7.99E-01	kg
Metalworking factory	4.58E-10	p
Outputs		
Heat, waste	3.87E-01	MJ

The CED method (Röhrlich et al., 2000) analyzes the environmental effects based on energy consumption in the form of fossil fuels, nuclear, renewables, and nonrenewables. Fig. 12.7 depicts the CED and CML, the LCA approaches. Overall, the major components of these methods are analyzed and compared here based on their respective impact categories to perform a detailed and complete life cycle impact analysis based on Li-ion and NiMH batteries.

12.4 Results

12.4.1 Impacts of Li-ion batteries

Results from the application of the Impact 2002 + method illustrate the impacts of Li-ion batteries in terms of 12 environmental indicators, as presented in Fig. 12.8. Li-ion batteries have high impacts on respiratory inorganics (1.476399 mPt), noncarcinogens (1.329504 mPt), nonrenewable energy (0.635 mPt), mineral extraction (0.05 mPt), and carcinogens (0.282865 mPt). Fig. 12.3 demonstrates the gas- and metal-based emissions from producing 1 kg of Li-ion batteries using the EcoPoints 97 method. As obtained from Fig. 12.9, two top-ranked gas- and metal-based emissions from Li-ion batteries are SO_x (3.28 kPt) and copper

Table 12.4 Life cycle inventory of NiMH battery.

Products	**1.00**	**kg**
Battery, NiMH, rechargeable, prismatic, at plant/GLO U/AusSD U		
Inputs		
Acrylic acid, at plant	1.34E-03	kg
Copper, primary, at refinery	3.62E-06	kg
Electricity, medium voltage, at grid	3.26E-01	kWh
Electricity, medium voltage, at grid	2.18E-01	kWh
Electrode, negative, Ni	3.59E-01	kg
Electrode, positive, LaNi5	3.35E-01	kg
Electrolyte, KOH, LiOH additive	8.22E-02	kg
Facilities precious metal refinery	1.65E-07	p
Heat, natural gas, at industrial furnace	8.35	MJ
Injection molding	1.18E-01	kg
Nickel, 99.5%	5.83E-02	kg
Polycarbonate	8.12E-02	kg
Polyethylene, LDPE, granulate	1.83E-02	kg
Polypropylene, granulate	1.85E-02	kg
Sheet rolling, copper	3.62E-06	kg
Sheet rolling, steel	1.05E-01	kg
Steel, low-alloyed	4.63E-02	kg
Transport, freight, rail	9.47E-02	tkm
Transport, lorry >16t	3.06E-02	tkm
Water, decarbonized	1.83E + 02	kg
Zinc coating	4.81E-05	m^2
Zinc, primary	3.62E-08	kg
Outputs		
Heat, waste	7.73E + 01	MJ
Disposal, hazardous waste, 25% water	8.84E-01	kg

(3.2 kPt), respectively. Fig. 12.10 compares the number of various types of fuel (fossil and renewable) consumption by Li-ion batteries applying the CED method. As depicted, Li-ion batteries consume fossil fuels the most (81.73753 MJ), followed by nuclear energy (13.40453 MJ) and renewable/water energy (6.77425 MJ).

12.4.2 Impacts of NiMH batteries

Results from the Impact 2002 + method illustrate the impacts of NiMH batteries based on several factors, as presented in Figs. 12.11 and 12.12. According to Fig. 12.11, NiMH batteries have a towering impact on

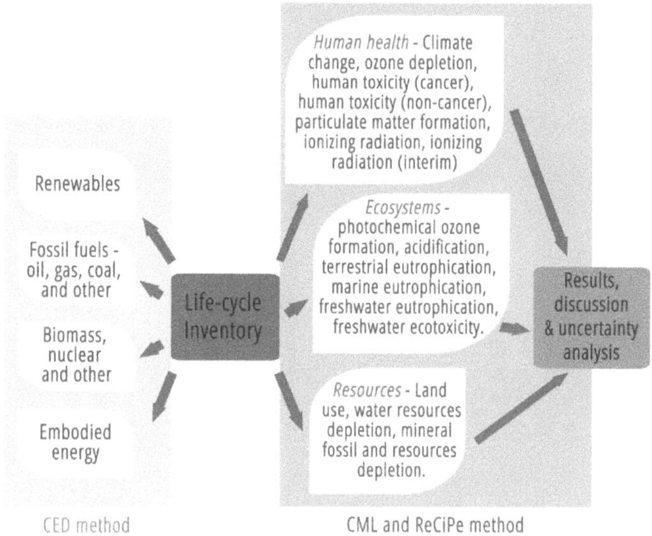

Figure 12.7 Schematic diagram of the CED and CML methods of LCA analysis. *From Parvez Mahmud, M.A., Huda, N., Farjana, S.H., Lang, C., 2019. A strategic impact assessment of hydropower plants in alpine and non-alpine areas of Europe, Applied Energy, 250, 198-214. https://doi.org/10.3390/batteries5010022.*

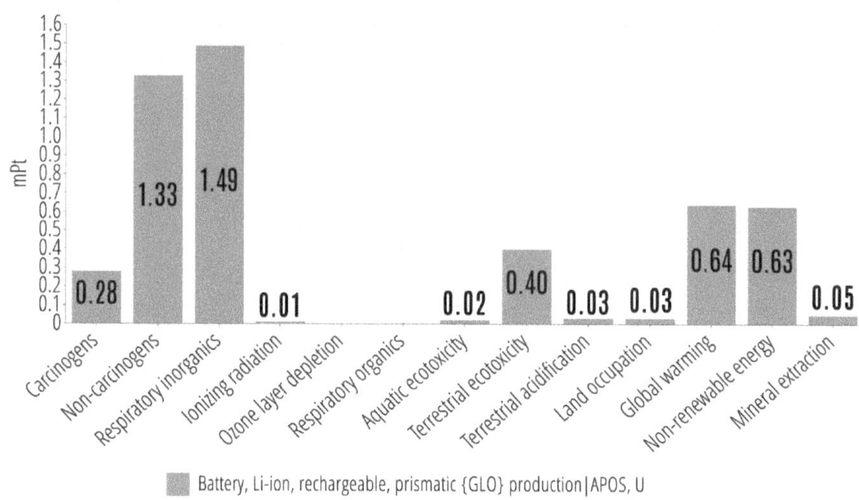

Figure 12.8 Comparison per impact indicator, using the Impact 2002 + method.

respiratory inorganics with a value of 10.75 mPt. Other considerable values from NiMH batteries comprise global warming (1.75 mPt), nonrenewable energies (1.5 mPt), and terrestrial ecotoxicity (0.75 mPt). From Fig. 12.12, it is depicted that NiMH batteries have a soaring impact on human health

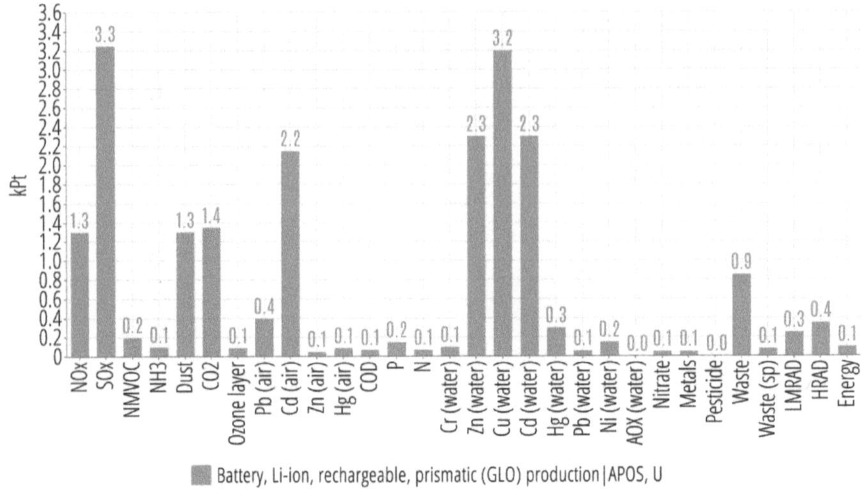

Figure 12.9 Comparison per impact indicator, using the EcoPoints 97 method. *From Parvez Mahmud, M.A., Huda, N., Farjana, S.H., Lang, C., 2019. A strategic impact assessment of hydropower plants in alpine and non-alpine areas of Europe, Applied Energy, 250, 198-214. https://doi.org/10.3390/batteries5010022.*

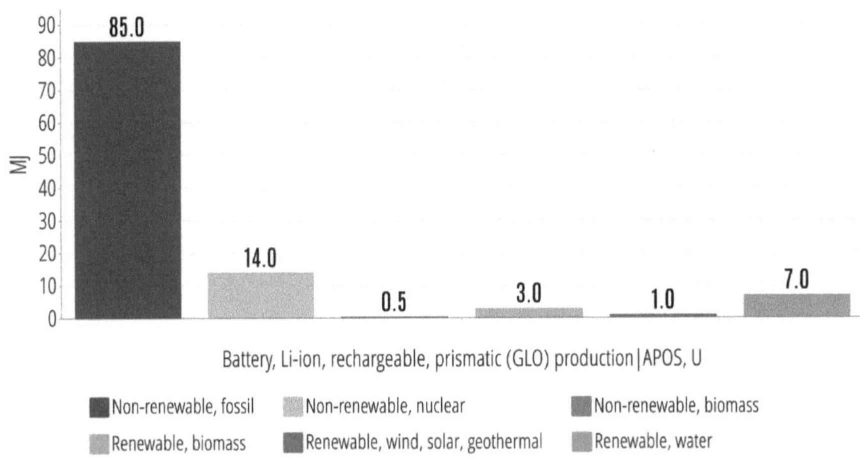

Figure 12.10 Comparison per impact indicator, using the CED method.

with a value of 11.2 mPt. The detrimental emissions to the environment from producing 1 kg of NiMH batteries are presented in Fig. 12.13 using the EcoPoints 97 method. As presented in the figure, the highest value of harmful environmental emission is from SO_x with around 58 kPt. Nevertheless, Fig. 12.14 reveals the consumption of various fuel types by

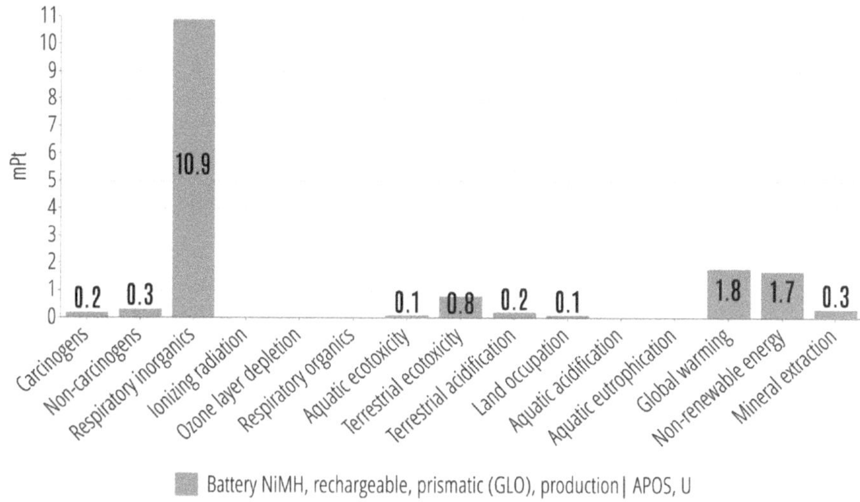

Figure 12.11 Comparison per impact indicator, using the Impact 2002 + method.

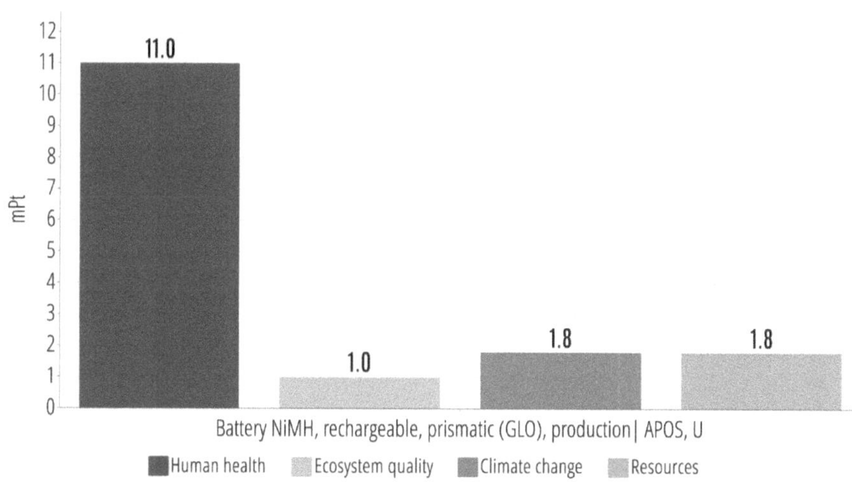

Figure 12.12 Comparison per impact indicator, using the Impact 2002 + method.

NiMH batteries obtained by the CED method. As depicted, NiMH consumes fossil fuel the most with a value of nearly 220 MJ followed by nuclear and renewable/water energy with values of 20 and 19.8 MJ, respectively.

12.4.3 Impacts of NaCl batteries

Results from the implication of the Impact 2002 + method in the case of NaCl batteries are illustrated in Figs. 12.15 and 12.16 based on several

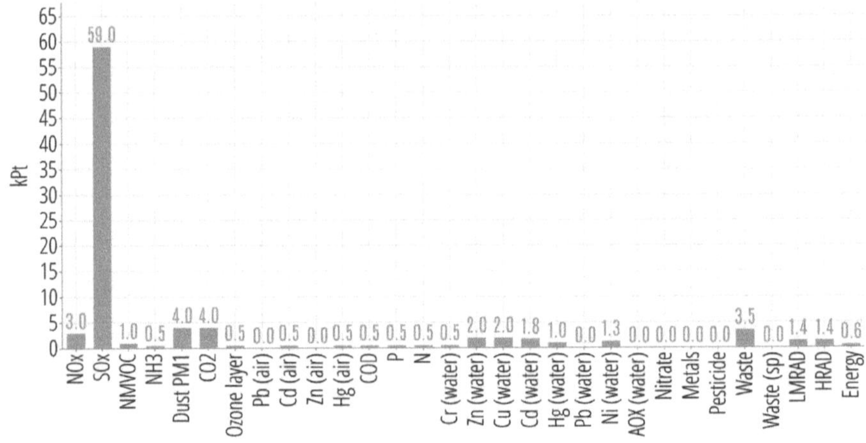

Figure 12.13 Comparison per impact indicator, using the EcoPoints 97 method. *From Parvez Mahmud, M.A., Huda, N., Farjana, S.H., Lang, C., 2019. A strategic impact assessment of hydropower plants in alpine and non-alpine areas of Europe, Applied Energy, 250, 198-214. https://doi.org/10.3390/batteries5010022.*

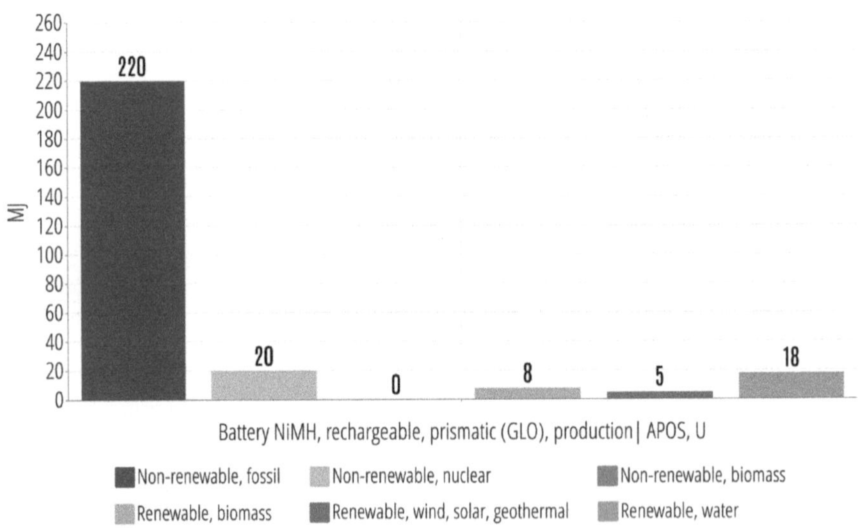

Figure 12.14 Comparison per impact indicator, using the CED methods.

impact indicators. As presented in Fig. 12.15, NaCl batteries have high impacts on respiratory inorganics with a value of around 4.78 mPt, whereas the adverse impact of NaCl batteries on human health occupies the value of 3.6×10^{-5} mPt, as depicted from Fig. 12.16. The

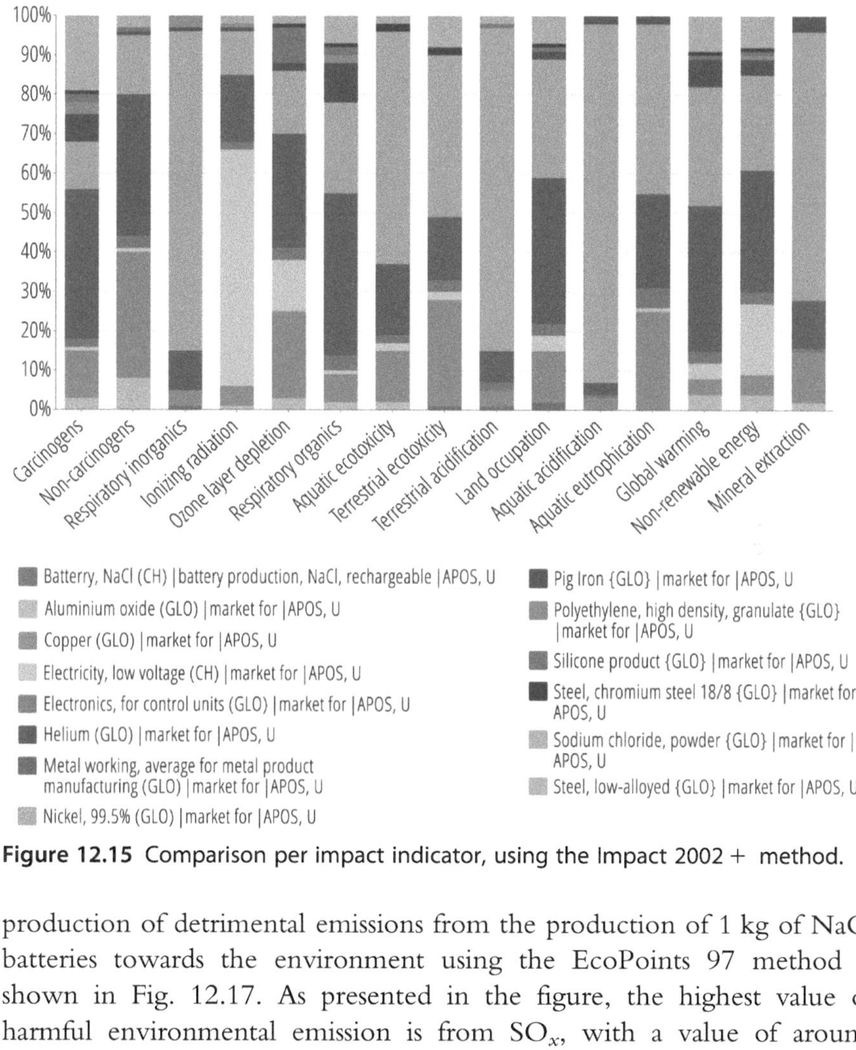

Figure 12.15 Comparison per impact indicator, using the Impact 2002 + method.

production of detrimental emissions from the production of 1 kg of NaCl batteries towards the environment using the EcoPoints 97 method is shown in Fig. 12.17. As presented in the figure, the highest value of harmful environmental emission is from SO_x, with a value of around 472 g. From Fig. 12.18, it can be observed that the production of NaCl consumes fossil fuels the most, with a value of around 71.6 MJ, obtained from the CED method.

12.5 Discussion

12.5.1 Impact outcome comparison

In this section, the outcomes of the environmental profile evaluation of Li-ion, NiMH, and NaCl batteries are accumulated for a proper comparison, as

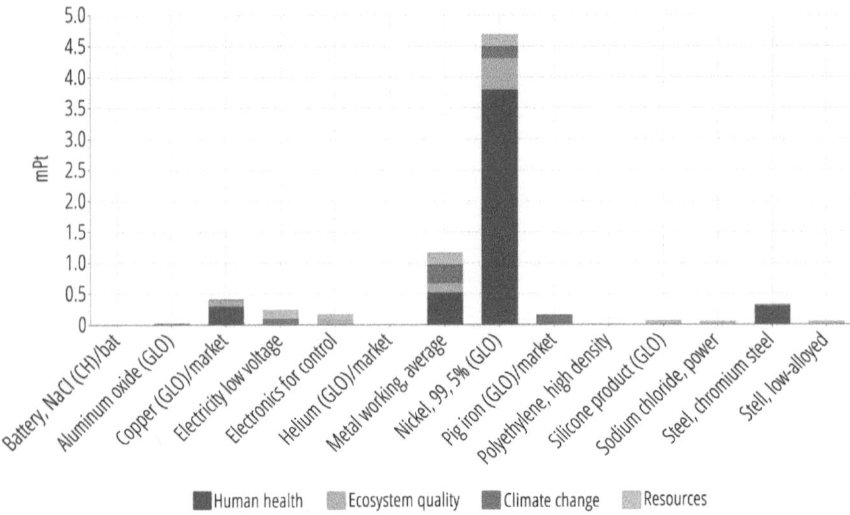

Figure 12.16 Comparison per impact indicator, using the Impact 2002 + method.

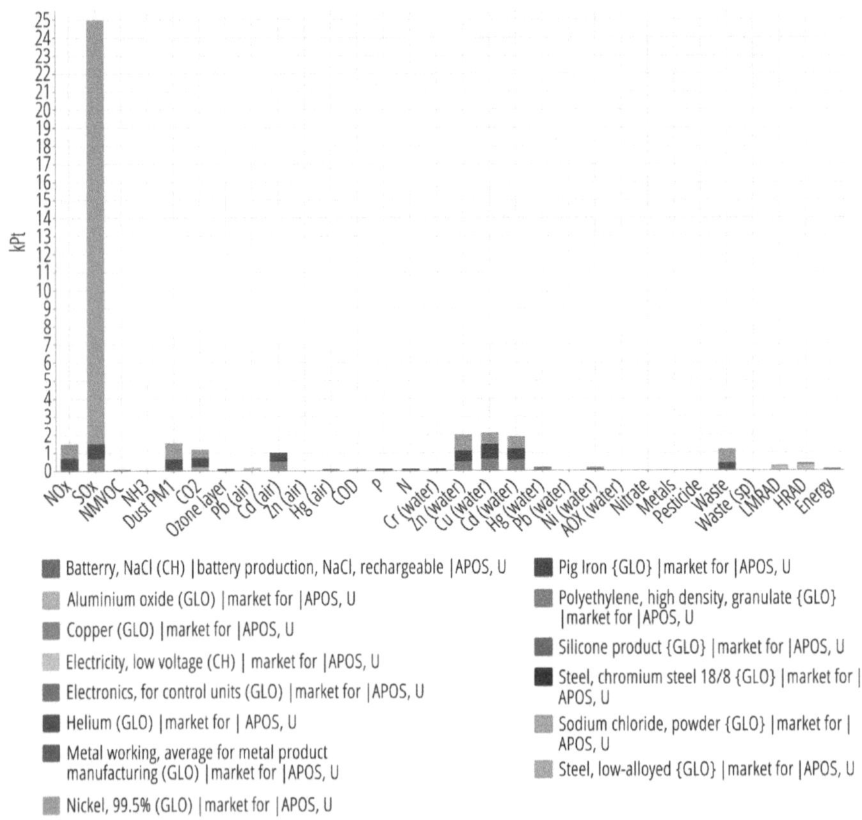

Figure 12.17 Comparison per impact indicator, using the EcoPoints 97 method.

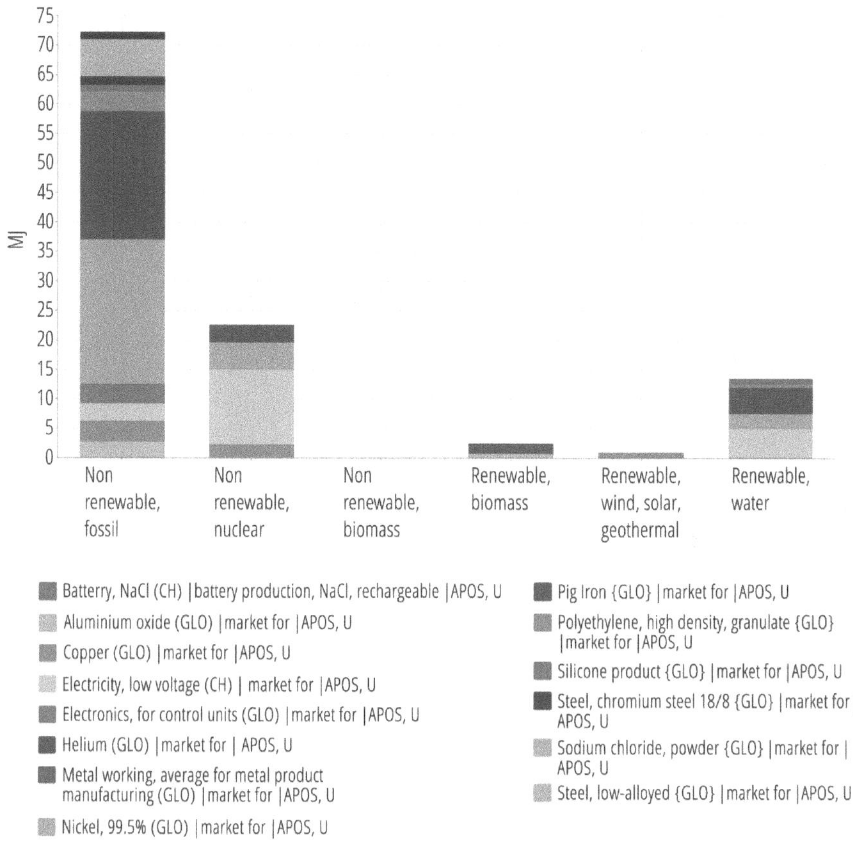

Figure 12.18 Comparison per impact indicator, using the CED method.

presented in the tables below. Among all three types of batteries, NiMH batteries have the highest impact on global warming, terrestrial ecotoxicity, and respiratory inorganics, whereas Li-ion batteries have the highest impact on carcinogens and noncarcinogens, as presented in Table 12.5. From Table 12.6, it can be observed that NiMH batteries contain the most harmful impacts on human health, followed by NaCl and Li-ion batteries. Also, NiMH batteries have the highest impacts on ecosystem quality and climate change compared to the other two types of batteries. As obtained from Table 12.7, the value of SO_x emission is highest for NaCl batteries at 472.1272 g, and the value of NO_x emission is highest for NiMH batteries with a value of 46.73259 g. Finally, Table 12.8 illustrates that Li-ion, NaCl, and NiMH batteries consume fossil fuels the most, followed by nuclear energy.

Table 12.5 Comparison per impact indicator using the Impact 2002 + method among Li-ion, NaCl, and NiMH batteries.

Impact category	Unit	Li-ion	NaCl	NiMH
Carcinogens	mPt	0.282865	0.071272	0.11361
Noncarcinogens	mPt	1.329504	0.214644	0.172049
Respiratory inorganics	mPt	1.476399	4.78029	10.7369
Ionizing radiation	mPt	0.002248	0.004149	0.003969
Ozone layer depletion	mPt	0.000183	0.000133	0.006397
Respiratory organics	mPt	0.000867	0.000854	0.001685
Aquatic ecotoxicity	mPt	0.008272	0.011281	0.021954
Terrestrial ecotoxicity	mPt	0.407027	0.508064	0.774249
Terrestrial acid/nutri	mPt	0.014207	0.047018	0.106351
Land occupation	mPt	0.013793	0.012905	0.024382
Global warming	mPt	0.636833	0.631338	1.726979

Table 12.6 Comparison per impact indicator using the Impact 2002 + method among Li-ion, NaCl, and NiMH batteries.

Impact category	Unit	Li-ion	NaCl	NiMH
Human health	DALY	2.19E-05	3.6E-05	7.83E-05
Ecosystem quality	PDF*m2*yr	6.072605	7.935161	12.69775
Climate change	kg CO_2 equiv.	6.30528	6.250876	17.0988
Resources	MJ primary	102.4367	107.333	260.8933

PDF*m2*yr, Potentialy disappeared fraction of species on a m2 during a year.

12.6 Limitations

In every LCA research, there are a few limitations. For example, a geographic location for data collection or a particular part of the life cycle instead of whole life is considered. However, the main limitations of this work are as follows:

- The environmental profiles of other battery storages except for Li-ion, NaCl, and NiMH have not been assessed and compared because of datasets' unavailability.
- The replacement of dangerous battery storage parts, elements, or machines has not been identified.
- The uncertainty analysis has not been conducted.
- The life cycle cost-benefit analysis has not been studied.

Table 12.7 Comparison per impact indicator using the EcoPoints 97 method among Li-ion, NaCl, and NiMH batteries.

Impact category	Unit	Li-ion	NaCl	NiMH
NO_x	g	19.2221	21.85741	46.73259
SO_x	g SO_2 equiv.	61.68981	472.1272	1098.069
NH3	g	1.440182	1.885402	3.362038
CO_2	g CO_2 equiv.	6680.584	6616.102	18117.49
N	g	0.831354	1.164624	1.432781
Nitrate (soil)	g	0.889376	0.688981	1.646231
P	g	0.057105	0.024821	0.061701
Pb (air)	g	0.1303	0.07245	0.028557
Cd (air)	g	0.017106	0.008184	0.001407
Zn (air)	g	0.055248	0.055534	0.084718
Hg (air)	g	0.000603	0.000607	0.000735
Cr (water)	g	0.138326	0.163374	0.21241
Cu (water)	g	2.655404	1.853189	1.883439
Ni (water)	g	0.592575	1.831099	3.192432

Table 12.8 Comparison per impact indicator using the CED method among Li-ion, NaCl, and NiMH batteries.

Impact category	Unit	Li-ion	NaCl	NiMH
Nonrenewable, fossil	MJ	81.73753	71.60173	218.5246
Nonrenewable, nuclear	MJ	13.40453	22.34603	19.62816
Nonrenewable, biomass	MJ	0.021481	0.013616	0.159694
Renewable, biomass	MJ	2.859748	2.894852	4.614767
Renewable, wind, solar, geothermal	MJ	0.675969	0.999048	1.480717
Renewable, water	MJ	6.77425	12.0941	15.39008

Future work should consider and resolve the above constraints to eliminate the environmental impacts of battery storage. The utilization of renewable energy sources instead of fossil-fuel-based power sources is inevitable for green battery production. For cleaner disposal and recycling of end-of-life battery parts, an appropriate measure should be carried out. Thus, the overall environmental impacts of battery storage could be avoided.

12.7 Conclusion

This research work applied LCA analysis to estimate and compare the environmental profiles of Li-ion, NaCl, and NiMH battery storage over the entire lifespan, from the extraction of raw material to the end-of-life disposal stages. Several well-constructed methods such as the Impact 2002 +, EcoPoints 97, and CED were employed in evaluating the impacts. The SimaPro software and Ecoinvent database were applied in performing the life cycle analysis (LCA). The essential stages included determining the LCA goal, building the LCI, and assessing and comparing the impacts. The mid-point impact indicators under consideration were carcinogens, noncarcinogens, respiratory inorganics, ionizing radiation, ozone layer depletion, aquatic ecotoxicity, land occupation, mineral extraction, nonrenewable energy, terrestrial ecotoxicity, respiratory organics, and global warming. The analyzed end-point impact indicators were human health, ecosystem quality, climate change, and resources. In addition, both GHG emissions and metal-based emissions were assessed for the storages of Li-ion, NaCl, and NiMH batteries. The results show Li-ion batteries produce a lower amount of GHGes as well as metals like carbon dioxide (8.27 kg CO_2 equiv.), nitrogen oxide (2.54 E-02 kg N_{20} equiv.), sulfur oxide (3.74 E-02 kg S20 equiv.), ammonia (9.53 E-04 kg), phosphorous (4.89 E-05 kg), lead (9.85 E-05 kg), cadmium (1.20 E-05 kg), zinc (6.15 E-05 kg), and mercury (5.19 E-07 kg) compared to NiMH batteries. Also, less amount of fossil fuels is required by the Li-ion batteries throughout their lifespan. Nevertheless, from the uncertainty analysis, it is depicted that the inventory datasets of the batteries are firm enough because of less diversity for most of the impact categories. All in all, Li-ion batteries supersede NiMH batteries in terms of environmental impacts with the following impact indicators such as abiotic depletion (5.23 E-05 MJ), acidification (5.49 E-02 kg SO_2 equiv.), human toxicity (7.38 kg 1,4-DB equiv.), ozone layer depletion(2.90 E-07 kg CFC-11 equiv.), freshwater aquatic ecotoxicity (1.60 E-01 kg 1,4-DB equiv.), marine aquatic ecotoxicity (8.62 E + 03 kg 14-DB equiv.), terrestrial ecotoxicity (2.63 E-02 kg 1,4-DB equiv.), and photochemical oxidation (2.70 E-03 kg C2H4 equiv.). These findings will certainly assist the scientists in keeping the environment away from the hazardous effects of batteries by providing a suitable environment-friendly substitute in the place of the impacting elements.

Nomenclature

Term	Description
kg CO_2 equiv.	kilograms of carbon dioxide equivalent
kg CFC-11 equiv.	kilograms of CFC-11 equivalent
kg Cl^- equiv.	kilograms of chlorine equivalent

kg Cd equiv.	kilograms of cadmium equivalent
kWh	kilowatt-hour
tkm	tonne-kilometer
kg N equiv.	kilograms of nitrogen equivalent
kg O_3 equiv.	kilograms of ozone equivalent
MJ LHD	mega joules of lower heating values
MJ HHD	mega joules of higher heating values
kg 1,4-DB equiv.	kilograms of 1,4 dichlorobenzene equivalent
kg C_2H_4 equiv.	kilograms of ethylene equivalent
kg PO_4 equiv.	kilograms of phosphate equivalent
kg SO_2 equiv.	kilograms of sulfur dioxide equivalent
kg NO_2	Kilograms of nitrogen dioxide equivalent

References

Audenaert, A., De Cleyn, S.H., Buyle, M., 2012. LCA of low-energy flats using the Eco-indicator 99 method: Impact of insulation materials. Energy and Buildings 47, 68–73.

Battery Association of Japan, 2018. BAJ battery sales statistics by volume and value. <http://www.baj.or.jp/e/> (accessed 20.01.20).

Curran, M.A., 2013. Life cycle assessment: a review of the methodology and its application to sustainability. Current Opinion in Chemical Engineering 2 (3), 273–277.

DOE Global Energy Storage Database, Office of electricity delivery & energy reliability, 2017. <https://sandia.gov/ess-ssl/gesdb/public/index.html> (accessed 20.01.2021).

Ellingsen, L.A.W., Majeau-Bettez, G., Singh, B., Srivastava, A.K., Valøen, L.O., Strømman, A.H., 2014. Life cycle assessment of a lithium-ion battery vehicle pack. Journal of Industrial Ecology 18 (1), 113–124.

Farjana, S.H., Huda, N., and Mahmud, M.P., Life-cycle environmental impact assessment of mineral industries. In IOP Conference Series: Materials Science and Engineering, IOP Publishing, 351(1), May 2018a, p. 012016.

Farjana, S.H., Huda, N., Mahmud, M.P., Lang, C., 2018b. Comparative life-cycle assessment of uranium extraction processes. Journal of Cleaner production 202, 666–683.

Farjana, S.H., Huda, N., Mahmud, M.P., Lang, C., 2018c. Towards sustainable TiO2 production: An investigation of environmental impacts of ilmenite and rutile processing routes in Australia. Journal of Cleaner Production 196, 1016–1025.

Farjana, S.H., Huda, N., Mahmud, M.P., Saidur, R., 2018d. Solar process heat in industrial systems—A global review. Renewable and Sustainable Energy Reviews 82, 2270–2286.

Finnveden, G., Hauschild, M.Z., Ekvall, T., Guinée, J., Heijungs, R., Hellweg, S., et al., 2009. Recent developments in life cycle assessment. Journal of Environmental Management 91 (1), 1–21.

Hao, H., Mu, Z., Jiang, S., Liu, Z., Zhao, F., 2017. GHG emissions from the production of lithium-ion batteries for electric vehicles in China. Sustainability 9 (4), 504.

Hischier R., Classen M., Lehmann M., and Scharnhorst W., 2007. Life cycle inventories of electric and electronic equipment: production, use and disposal. Swiss Centre for Life Cycle Inventories, UK.

IEA, 2016. World Energy Outlook 2016, International energy agency, Paris,<https://doi.org/10.1787/weo-2016-en> (accessed 20.01.2021).

Innocenzi, V., Ippolito, N.M., De Michelis, I., Prisciandaro, M., Medici, F., Vegliò, F., 2017. A review of the processes and lab-scale techniques for the treatment of spent rechargeable NiMH batteries. Journal of Power Sources 362, 202–218.

Lewandowska, A., Matuszak-Flejszman, A., Joachimiak, K., Ciroth, A., 2011. Environmental life cycle assessment (LCA) as a tool for identification and assessment of environmental aspects in environmental management systems (EMS). The International Journal of Life Cycle Assessment 16 (3), 247–257.

Liang, Y., Su, J., Xi, B., Yu, Y., Ji, D., Sun, Y., et al., 2017. Life cycle assessment of lithium-ion batteries for greenhouse gas emissions. Resources, conservation and recycling 117, 285–293.

Ligthart, T.N., Jongbloed, R.H., Tamis, J.E., 2010. A method for improving Centre for Environmental Studies (CML) characterisation factors for metal (eco) toxicity—the case of zinc gutters and downpipes. The. International Journal of Life Cycle Assessment 15 (8), 745–756.

Mahmud, M.A.P., Farjana, S.H., 2012. Design and construction of refrigerant charge level detecting device in HVAC/R system with microcontroller. International Journal of Engineering and Advanced Technology (IJEAT) 1, 309–314.

Mahmud, M.P., Lee, J., Kim, G., Lim, H., Choi, K.B., 2016. Improving the surface charge density of a contact-separation-based triboelectric nanogenerator by modifying the surface morphology. Microelectronic Engineering 159, 102–107.

Mahmud, M.A., Huda, N., Farjana, S.H., Lang, C., 2018a. Environmental impacts of solar-photovoltaic and solar-thermal systems with life-cycle assessment. Energies 11 (9), 2346.

Mahmud, M.A.P., Huda, N., Farjana, S.H. and Lang, C., Environmental sustainability assessment of hydropower plant in Europe using life cycle assessment. In IOP conference series: Materials science and engineering, IOP Publishing, 351(1) May, 2018b, p. 012006.

Mahmud, M.P., Huda, N., Farjana, S.H. and Lang, C., Environmental life-cycle assessment and techno-economic analysis of photovoltaic (PV) and photovoltaic/thermal (PV/T) systems. In 2018 IEEE international conference on environment and electrical engineering and 2018 IEEE industrial and commercial power systems Europe (EEEIC/I&CPS Europe), IEEE, June, 2018c, pp. 1–5.

Mahmud, M.P., Huda, N., Farjana, S.H. and Lang, C., Environmental profile evaluations of piezoelectric polymers using life cycle assessment. In IOP Conference Series: Earth and Environmental Science, IOP Publishing,154 (1) May, 2018d, p. 012017.

Marsmann, M., Schiburr, A., 1993. Databases, software and LCA applications. Journal of Cleaner Production 1 (3), 206–207.

Meng, T., Young, K.H., Wong, D.F., Nei, J., 2017. Ionic liquid-based non-aqueous electrolytes for nickel/metal hydride batteries. Batteries 3 (1), 4.

Minx, J.C., Callaghan, M., Lamb, W.F., Garard, J., Edenhofer, O., 2017. Learning about climate change solutions in the IPCC and beyond. Environmental Science & Policy 77, 252–259.

Notter, D.A., Gauch, M., Widmer, R., Wager, P., Stamp, A., Zah, R., et al., 2010. Contribution of Li-ion batteries to the environmental impact of electric vehicles. Environmental Science & Technology 44 (17), 6550–6556.

Parvez Mahmud, M.A., Huda, N., Farjana, S.H., Asadnia, M., Lang, C., 2018. Recent advances in nanogenerator-driven self-powered implantable biomedical devices. Advanced Energy Materials 8 (2), 1701210.

Pascual-González, J., Guillén-Gosálbez, G., Mateo-Sanz, J.M., Jiménez-Esteller, L., 2016. Statistical analysis of the ecoinvent database to uncover relationships between life cycle impact assessment metrics. Journal of Cleaner Production 112, 359–368.

Pesso, C., 1993. Life cycle methods and applications: issues and perspectives. Journal of Cleaner Production 1 (3–4), 139–142.

Pryshlakivsky, J., Searcy, C., 2013. Fifteen years of ISO 14040: a review. Journal of Cleaner Production 57, 115–123.

Röhrlich, M., Mistry, M., Martens, P.N., Buntenbach, S., Ruhrberg, M., Dienhart, M., et al., 2000. A method to calculate the cumulative energy demand (CED) of lignite extraction. The International Journal of Life Cycle Assessment 5 (6), 369–373.

Stavropoulos, P., Giannoulis, C., Papacharalampopoulos, A., Foteinopoulos, P., Chryssolouris, G., 2016. Life cycle analysis: comparison between different methods and optimization challenges. Procedia CIRP 41, 626–631.

Total Battery Production Statistics of Japan 2017, Machinery statistics released by the ministry of economy, trade and industry. <https://www.baj.or.jp/e/statistics/01.html> (accessed 20.01.20).

Židonienė, S., Kruopienė, J., 2015. Life cycle assessment in environmental impact assessments of industrial projects: towards the improvement. Journal of Cleaner Production 106, 533–540.

SECTION D

CHAPTER 13

Environmental assessment of renewable energy and storage technologies: future challenges

Panagiota Konatzii and Paris A. Fokaides
School of Engineering, Frederick University, Nicosia, Cyprus

The purpose of this edited volume was to compile the existing knowledge in the field of environmental assessment of sustainable and renewable technologies for producing energy and energy storage with the use of life cycle assessment (LCA). This volume hosts 11 individual chapters, which cover the main topics of renewable energy production and energy storage that are currently at the forefront of the energy sector. The environmental assessment of photovoltaic (PV) systems, solar thermal systems, wind turbines, thermochemical, biochemical, and mechanical conversion routes of biomass to biofuels, geothermal energy, hydroelectricity, hydrogen, and energy storage with compressed air and with batteries were presented. Some of these technologies are more mature, resulting in numerous environmental assessment studies, allowing for the extraction of solid conclusions with regard to the research trends in the field. In contrast, other topics such as hydrogen are still under development; to this end, the environmental assessment is of different nature and content.

The main findings of this edited volume concerning the future challenges of the field could be summarized as follows:

- With regard to the environmental assessment of "PV systems", it was deduced that LCA studies in this field mainly focus on mono-crystalline and poly-crystalline Si panels and from nonsilicone-based panels cadmium–telluride (CdTe), copper–indium–selenide (CIS), or copper–indium–gallium–diselenide (CIGS) cells. The increased concern is shown in the majority of the studies in the cradle-to-gate approach. This is due to the PV elements construction related to high energy demand resulting in an increase of CO_2 emissions as well as

Environmental Assessment of Renewable Energy Conversion Technologies
DOI: https://doi.org/10.1016/B978-0-12-817111-0.00012-7

greenhouse gas (GHG) emissions. Future research targets in this field will be aimed to improve the environmental impacts of the materials used in PV cells focusing on high technical efficiency energy systems and cost-effective solutions.

- As far as the "solar thermal systems" are concerned, it was revealed that a comprehensive analysis of the system life cycle, including the often-ignored disposal or decommissioning stage, and all environmental impacts related to the studied systems, for example, not just focusing on climate changes or energy-related issues, should be better addressed in future research.
- The environmental assessment of "wind energy technologies" undertaken in this work has led to the identification of the number of challenges for the LCA practitioners and the scientific community of the relevant field. Some of the challenges identified are horizontal, shared challenges common across almost any type of product or product system assessed using LCA methodologies, whereas others are technology-specific. Common challenges across several fields of LCA application include the choice of assumptions, methodologies, and the phase of data collection for the development of life cycle inventories (LCIs). In the case of technology-specific challenges, considerable concerns arise from the technological, geographical, and temporal scope of LCA studies of wind turbines. The most contributing factor for these challenges is the nature of the technology itself, which is very site-depended, as is the case for all renewable energy systems. LCA methodologies in the case of wind turbines have also been criticized for not addressing major environmental issues; additional environmental burdens that are relevant to wind power include land use and noise. However, in terms of land usage requirements and the amount of energy to be extracted in relation to the land area, wind turbines have been proven to perform well in comparison to conventional power plants, overruling the claims for excessive land use by wind energy.
- Concerning "thermochemical pathways for converting biomass to biofuels", environmental performance evaluation is complicated; differentiations in feedstock, conversion pathways, intended application, and methodological choices lead to a wide range of results in terms of net energy balances and net GHG emissions. Concerning methodological choices in the goal and scope definition, the cradle-to-grave approach was found advantageous over cradle-to-gate. Regarding methodological choices in the LCI stage,

information on data sources is usually reported, high usage of scientific literature and LCA databases was found for both foreground and background processes. However, regional relevance, data age, and assumptions affect the quality and, therefore, the overall result. Concerning methodological choices in the life cycle impact assessment stage, most studies address the GWP, followed by acidification potential and eutrophication potential. Energy evaluation is also included, but the various indicators hinder comparability. Land-use change is not widely studied in the LCA of the thermochemical processing of biomass mainly due to the lack of an established and widely accepted method.

- With regard to "biochemical conversion routes of biomass to deliver biogas", each step generates an environmental impact, which should be aggregated and considered along the environmental impacts created in all life cycle stages of the product. LCA was found to enable fair comparisons between biofuels and conventional fuels, considering a wide range of impact categories, such as global warming potential, abiotic depletion, human toxicity potential, and land occupation. The product environmental footprint (PEF) provides more specific requirements for performing an LCA for individual product categories. No PEF category rules for biofuels and fuels, in general, were documented, and their development to support LCA studies of biofuels is highly encouraged.
- LCA studies of "solid biofuel conversion systems", including biomass densification, were found more than necessary to identify the hotspots with regard to their environmental impact. Although a standardized methodology guides the framework of LCA studies, the variations in the definition of system boundaries, the uncertainty of data in many cases, as well as the assumptions applied always make the comparison of different studies quite challenging and sometimes irrelevant. The results of previous LCA studies in biomass densification highlighted the significant impact of the pelleting/briquetting stage in such systems, as well as the contribution of transportation to their environmental footprint.
- As far as "geothermal energy" is concerned, environmental assessment studies are typically compared with other power generation technologies, whether renewables or not. For example, it might be instructive to compare the environmental footprint of geothermal energy production against that connected with onshore/offshore wind, solar PV, boiling and pressure water nuclear reactors, and fossil fuel-powered

plants for the provision of electricity. Geothermal energy can also be compared with other technologies to provide thermal energy. For instance, it can be compared with electric or natural gas boilers or heat pumps for the provision of space heating.

- The assessment of "hydroelectric power technologies" revealed that alpine zone-based hydropower plants provide a superior environmental profile compared to nonalpine zones. In addition, the assessment of the GHG emissions was provided that hydropower plants in alpine zones emit about 10 times lower methane gas in comparison to nonalpine zone-based power plants. All in all, it is proposed that in order to elevate the generation of clean hydropower in the future, further research is needed to be more focused on detecting the most dangerous materials during plant installation and replacing them with matching ecologically superior options.
- The assessment of "hydrogen LCA" defined the various social risks associated with hydrogen production. These social risks can be interpreted as adverse external effects of hydrogen production at the three European production sites. This social risk model thus allows the generation of detailed data about the social effects of the current—as well as changing—energy production and consumption patterns, as exemplarily shown for the case of moving towards a hydrogen economy in the given case.
- With regard to storage technologies, the evaluation of the environmental footprints of "compressed air energy storage (CAES)" systems revealed that there are a few LCAs of conventional-CAES and adiabatic-CAES. However, no LCA was found for the emerging technology, liquid air energy storage (LAES). A LAES system uses electricity to liquefy air that can be stored in a well-insulated tank, and it can be built anywhere. An in-depth analysis of its environmental performance is required, and future LCAs should include it. Most of the studies reviewed here address the GHG emissions only. Other environmental impacts, that is, acidification, land use, water use, freshwater ecotoxicity, freshwater eutrophication, human toxicity, etc. are not well covered.
- Finally, as far as the environmental assessment of battery storage is concerned, an LCA analysis is presented to estimate and compare the environmental profiles of Li-Ion, NaCl, and NiMH battery storages over the entire lifespan, from the extraction of raw material to end-of-life

disposal stages. The findings of this assessment will certainly assist the scientists in keeping the environment away from the hazardous effects of batteries by providing a suitable environment-friendly substitute in the place of the impacting elements.

Index

Note: Page numbers followed by "*f*" and "*t*" refer to figures and tables, respectively.

S

T

U

V

W

FSC
www.fsc.org
MIX
Papier | Fördert gute Waldnutzung
FSC® C083411

Kolibri 360 GmbH
Ferdinand-Jühlke-Straße 7
99095 Erfurt, Deutschland
produktsicherheit@kolibri360.de